Hard-Won Lessons for Founders & Engineers

DECODING ROBOTS

How to Build and Launch Robots in the Real World

By Prathamesh Khedekar

DECODING ROBOTS

preserve confidentiality while maintaining the integrity of the lessons.

Image credits and third-party attributions appear in a dedicated section of this book.

ISBN (Paperback): 979-8-9942966-0-8

ISBN (Hardcover): 979-8-9942966-1-5

Library of Congress Control Number: 2026900872

Published by Boring Sage
Lake Forest, Illinois
www.boringsage.com

Printed in the United States of America

Dedication

To my Aae, Dipti Khedekar,

who devoted her life to my education.

To my Baba, sister, wife, and family,

for their unwavering support throughout this journey.

Why I Wrote This Book

The conference room was silent. Our robot had just completed its demo—every sensor calibrated to the teeth, every line of code executed perfectly, every use case that mattered to this client demonstrated with military-grade precision. The client nodded. For a moment, it felt like our robot had done its job.

Then came the words no engineer or founder wants to hear:

"We've decided not to move forward at this time."

That moment stayed with me. We as a team, prepared for this demo for months, only to get slammed with another "Not this time." This and many moments like these made me realize that building robots is brutally hard. Building a *company* around robots is even harder.

99.99% of robotics startups die within a few years of launch. The mortality rate here is higher than that of some of the deadliest diseases on the planet. And yet, every year, thousands of talented engineers and founders enter this arena, unaware of the traps that routinely catapult even the most well-funded and talented robotics teams into the valley of death.

I've seen engineers spend months perfecting a demo that never turned into a deal. Founders bet years of their lives on a prototype, only to walk away with a paper full of wisdom for the next generation of builders. And clients who once loved what you built—until the next meeting, when you realize the doors are closed forever and they are not proceeding with your robot-led service. This book is for you — the builders, founders, and engineers trying to make robots work in the real world.

Around 2018, after having designed mission-critical systems for government agencies in the U.S., U.K., and the Netherlands during my time at Motorola, I became deeply interested in robotics—reading voraciously, building 2WD robots, studying ROS (Robot Operating System), and exploring every corner of this field.

During that time, I met a few inspiring founders of a Silicon Valley robotics startup and decided to join their team. Over the next few years, I helped launch our robot-led safety and security service for a wide array of clients ranging from startups to Fortune 100 giants across the U.S., U.K., and Australia. From running client demos to enabling large-scale rollouts, and leading intense engineering jam sessions to resolve a wide range of technical issues—debugging Wi-Fi dead zones, tuning sensor modules, fixing ROS maps, and integrating elevators with robots—I had a front-row seat to every challenge. This experience made me realize that running a robotics startup is brutally hard.

We made plenty of mistakes—some avoidable, some inevitable. This book is an attempt to capture those lessons for the next-generation robotics founders, engineers, and investors. People who are brave enough to build robots for the real world.

You'll get a firsthand look at what it truly takes to build and launch robots in the real world, and the risks that come with it. From engineering to sales, and from marketing to field operations, these stories uncover the galactic traps that can catapult even the most well-funded and talented robotics teams into the valley of death. In addition to these lessons, you'll also learn the exact approach adopted by companies like *Amazon, Boston Dynamics, and Tesla* to bring modern robots and humanoids to life—and to tackle the risks that come with them.

If I could hand my past self a manual on Day 01 of my robotics journey in 2018, this book would be it—a collection of lessons earned at an extreme cost of time, capital, and energy.

Whether you're a founder, an engineer, or an investor, my hope is simple: that these stories help you identify these galactic traps in time, address them proactively, and build robots that create lasting value for your customers, your teams, and your investors — with fewer scars than we carried.

How To Use This Book

This book has been divided into five core sections:

Section I gives you an overview of how a life in a chawl of Mumbai prepares one to thrive in the startup world. These are the lessons that I have carried with me to this day.

Section II covers my own experiences from the field—running demos for Fortune 100 companies and mid-sized clients across the globe, and the lessons embedded in those encounters.

Section III covers the foundations of modern robots, especially humanoids, and Tesla's approach to building humanoids for the automotive manufacturing sector.

Section IV sheds light on lessons you will learn from the exact approach adopted by leading robotics companies such as Amazon and Boston Dynamics to bring modern robots to life and address the corresponding challenges faced by clients in the warehousing, logistics, and manufacturing sectors.

Sections V and VI serve as deep dives into lessons and frameworks that you, as an engineer or founder respectively, must keep in mind as you build your career or startup in this domain.

Section VII outlines the next steps you can take to apply the lessons and frameworks from this book in your journey as a founder or an engineer.

So, how should you approach this book?

If You're an Engineer Looking to Break into Robotics

If you are an engineer looking to break into this field, first develop a deep understanding of the risks and challenges you will encounter in the field as you begin working with clients. From engineering to sales, and from demos to field deployments, you should develop a strong hold on lessons from the field covered in Sections I and II. Then choose your class of robots—humanoids, mobile robots, or robot dogs—and study the corresponding engineering playbooks decoded in Sections III and IV. Here, you will develop a deep understanding of how leading companies such as Tesla, Amazon, and Boston Dynamics are building next-generation robots and, most importantly, how they address the core bottlenecks inherent to each class of robot from both a technical and customer-adoption perspective.

Last but not least, read Section V & VII, and deeply internalize both the challenges you will encounter as you begin your job search and the step-by-step framework you will need to navigate these traps and secure a strong position in the industry.

If You're a Founder Building (or Planning to Build) a Robotics Startup

If you are a founder looking to build your startup in this arena—or already building one—Sections I, II, VI, and VII are absolute musts. From engineering to sales, and from marketing to customer psychology, they cover the risks and challenges that can make or break your company. The rest of the book is equally valuable, but if you are short on time, these sections illuminate the most lethal risks on your path and equip you with the corresponding frameworks you will need to address them proactively and maximize your chances of building a solid robotics company.

While it is impossible to capture every lesson that I have learned in this field into a single volume, I hope this book equips you with the most important ones and prepares you for the Ironman ahead. Most importantly, I hope it helps you navigate the toughest parts of this journey with fewer scars and a far stronger foundation than I had when I started.

Table of Contents

Part I — Origins & Early Lessons

1. From the Chawls of Mumbai to Silicon Valley 1
2. Silicon Valley & the Evolution of Robots 13
3. The Art of Building Robots in Silicon Valley 21

Part II — Building, Selling & Scaling Robots

4. Robotics Startups — Lessons You Learn the Hard Way 27
5. When Apple Said No to Our Robot 39
6. Why Enterprise Deals Live or Die on Security 51
7. Meeting with a Cloud Giant at a Robotics Conference.........63
8. How a Strategic Partner 10x'd Our Growth 71

Part III — The Era of Humanoids

9. The Era of Humanoids 81
10. Humanoid Foundations — Core Systems & Principles 85
11. Decoding Tesla's Approach to Building Humanoids 105

Part IV — Decoding Big Tech's Robot Playbook

12. Decoding Amazon's Approach to Building Robots.............141
13. Amazon's One-Million-Robot Traffic-Jam Problem 153
14. Decoding Amazon's Human–Robot Interaction Systems...169
15. Decoding Boston Dynamics' Approach to Building Spot....177
16. Decoding Boston Dynamics' Approach to Building Stretch 199
17. How DHL & Maersk Are Adopting the Stretch Robot Arm .209

Part V — Pointers for Robotics Engineers

18. Career Traps to Avoid in Robotics 215
19. The Job Search Traps No One Warns You About 225

20. The Boring Sage Method for Landing a Job in Robotics....237

Part VI — Hard Lessons Every Robotics Founder Should Know

21. Fundraising and Finance — Capital and Runway..............261
22. Engineering and Talent — Product and Team...................275
23. Sales and Marketing — Value and Trust............................299
24. Operations — Execution at Scale 317

Part VII — Next Steps

25. How Founders & Engineers Can Apply This Book.............343

Image Credits

Notes on Sources, Author Experience, and Educational Use

References

About the Author

Chapter 01: From the Chawls of Mumbai to Silicon Valley

When you spend 22 years of your life in chawls of Mumbai, you develop a deep appreciation for each moment in your life: each kernel of food on your plate, each drop of pure water that helps you quench your thirst, and each square foot of space that allows you to a good night's sleep. For chawls are anything but a slightly better version of slums, nothing more than that. And this is where my journey to Silicon Valley began. This was the battleground where I learned some of the most important lessons in endurance and perseverance — lessons that became my operating protocol and helped me navigate the realities of the startup world.

So, what is life in a chawl really like, and why does it turn out to be such good training for building a startup? To answer that question, let me take you through my time in the chawl — the 22 years that prepared me for the storms I would later face in the Valley.

Life Inside the Chawls of Mumbai

Chawls have been at the heart of Mumbai for ages. They were developed for the factory workers who migrated from the villages to the cities in search of better jobs. I was born in one such chawl located in Parel, Mumbai. Growing up, I spent most of my time with my mother, who put deep emphasis on education and discipline. This naturally meant that books and local newspapers became my long-term friends. I am grateful we could afford them.

Most families in chawl barely had enough money to put food on their plate and shelter on their head; there wasn't much left for education. This meant most young kids in the chawl didn't go to school. Many had to work during the daytime, so education was not an option for these young minds. As a fourth grader in the late 1990s, when I would walk back home and play with the kids in the chawl in the evening, this thought pierced through my heart every single day. On one hand, I was happy that I had friends to play with; on the other, I was deeply saddened at the same time whenever I saw those who weren't given the opportunity to study. Gradually, I began developing a deep sense of purpose—a mission in life.

This brings us to the first lesson I learned in the chawl — one that I carry with me to this day.

Lesson 1: To Face Life & Startups, You Need a Deep, Heart-Piercing Mission

Just as a child growing up in a chawl needs a deep, heart-piercing mission to rise above these circumstances, a founder needs a mission to survive the chaos of the startup world. Your mission should run deep in your arteries and veins, so deep that even if you were to stand in front of death, the gods of death should give you more time to live and complete your mission. That should be the calibre of your mission and your belief in it.

It is this mission that will keep you going when your product fails, when you receive a series of "nos" from investors, and when you lose your top employees at the worst possible time, leaving you stranded in this ocean of fear and worry. Always, keep your eyes on your mission — it is the only constant that will help you withstand the storms in both life and in startups.

With this lesson etched into my heart, it started becoming clear as to why my parents were working extremely hard to afford me and my sister the quality of education that they could, even knowing the chances of us solving this brutal puzzle of life, trenched in the showers of poverty and hopelessness, were next to zero. After all, how do you expect one to study in a 120 sq. ft room that serves as hall, bedroom, study room, shower room, and kitchen? All in one 120 sq ft. room.

Space was not the only constraint.

The sounds of more than 300 tenants packed in a space half the size of a football field were enough to keep you awake even at 3:00 a.m. Then there are the sounds of Mumbai: the perennial horns of cars, the calls of street vendors, the sirens of ambulances emerging from the hospital located barely a few feet across from the chawl — all resulting in chaos. Beyond space and sound, water was a scarce resource. Chawls are the lowest in the order of priority when it comes to expecting a daily supply of tap water, let alone pure water. If that's not enough, these chawls are more than 100 years old, and so, come monsoon, we had waterfalls in our room at least 10 times per season. Post-monsoon, there are festivals and marriage celebrations on the streets in neighboring areas — and that meant more sound. Loudspeakers were a norm.

How do you study in such an environment? How do you expect one to study?

This brings us to the second lesson I learned in the chawl: focus.

Lesson 2: Deep Inward Focus Is the Only Antidote to Chaos

Just as a child in a chawl must develop an almost surgical level of inward focus to study through noise, scarcity, and constant disruption, a founder must cultivate the same inner stillness to survive the chaos that comes with the startup life.

Focus is your only survival tool. You need to develop a strong shield against this chaos. You may ask: what kind of chaos does this translate to in startups? Well, in a startup, this noise takes a different form: product failures at moments that matter most to your customer, customer escalations, emotional storms that rage through your team, funding deadlines, and, above all, the relentless mental pressure to keep the company alive that destabilizes the brain of even the most disciplined founders and steers them away from their vision.

To operate in environments like these, you need deep focus. Without it, you will drown in this noise. With it, you can think, build, prioritise, and make decisions that matter—decisions that will help you and your team stay focused in the midst of this chaos. Deep inward focus is the only antidote you have at your disposal in this case.

So, how does a child in a chawl develop focus amidst all this chaos? It takes time. It took me time to come to terms with my reality and to focus on what truly mattered at the end of the day. I cried many times, and each time my mother held me in her arms. She was loving and caring in every one of those moments. Through her demeanour and actions, she persuaded us to stay committed to our mission, no matter the circumstances. She helped us develop a laser focus on that mission. It was in this chawl that I realized our only way out was education.

That is where my sister and I put our heads down and studied — 365 days a year, for 22 years straight, in a 120 sq. ft. room in a chawl. That, for us, was home for 22 years and the training ground that prepared us to face both life and startups. It was during this time in our lives that we learnt that the ability to endure pain over a long stretch of time is a mental superpower — one that has the potential to withstand any adversity that comes your way.

This realization helped us learn our third major lesson in life: endurance — a lesson I carried with me into the startup world.

Lesson 3: Endure More Than Anyone Else

Just as a child in a chawl has to endure all the pain that comes through chaos, scarcity, and adversity, a founder has to endure all the pain that comes along the way. No exceptions. Received a rejection from the fifth investor? No problem. Endure. Lost your only customer who carried 50 per cent of the revenue for your startup? No problem. Endure. Faced a personal setback while building your company? No problem. Endure. The list is almost infinite.

In the end, your chances of making it through this journey successfully are less than 0.01 per cent — both in the chawl and in the startup world — and, therefore, there is no alternative to endurance, or, put simply, your ability to take pain on a recurring basis. So, take it with a smile on your face.

Endurance soon became our first and last hope. With that hope, and the sheer grit that came with it, after 12 years of hard work, around 2008, I got accepted into an engineering program at a respected college under University of Mumbai, and completed my undergrad, ranking in the top 0.1% in the university. That was roughly 45+ courses in a matter of 48 months and 8 semesters, retaining a top-3 rank all throughout those semesters. We were still living in a chawl in a 120 sq ft. room. This meant each day was just a workout with books in a cramped space for those 48 months. This was a painful journey — physically and mentally — but it was also the one that reinforced the need to develop long-term endurance in life.

During this time, the only thing that helped me survive was the unwavering support of my parents and their deep love towards us. At the end of this four-year period of engineering, I managed to get a decent score in GRE, and, looking at my overall contributions and performance, I got admits from multiple universities, including the University of Southern California (USC). I accepted it, but now I was required to demonstrate that I had a substantial amount of funds to finance my master's degree at this top private university. My reality? I was nowhere even close to 30 per cent of what I had to show on paper in the next thirty days. If I failed, I would lose this and several other admits I had secured.

This brings us to the next lesson — a lesson every founder must learn early on about funding.

Lesson 4: Scarcity Is a Founder's Best Fundraising Mentor

So, around this time, I had the admit in hand but less than 30 per cent of the funds required to pursue my education abroad, and less than thirty days to raise those funds. This scenario is no different from what early-stage founders face. You will encounter many rejections, because most people will not believe in you, and there is no way around this except to learn from each of those rejections and know when and how to find an alternative route to finance your startup. There are multiple ways to do this outside conventional funding rounds: there are angels, self-funded routes, accelerators — there is no shortage here. But you must show proof of execution with results.

In my case, I had secured an admit from a top 1 per cent university after studying for twenty-two years in a chawl. That, in itself, was reasonable proof. And if that was not enough, then it became clear that I had to find another way.

Given these conditions, my father and I decided to visit every bank we could in and around Mumbai to secure an education loan. We started a drill: each day we would leave home at around 8.00 a.m. and visit local branches of both regional and international banks. This routine continued for the next three weeks. It was the peak of summer in Mumbai, and within the first two hours of each day we would both be drenched in sweat. We would take a short break at a railway station or a bus stop, watching the crowds of Mumbai rush by, and then walk to the next bank. At the end of this exercise that lasted for three weeks, we returned home with rejections — not one, but more than twenty-one.

These rejections helped me learn my next lesson on life and startups.

Lesson 5: Pain and Rejections Become Your Recurring Revenue — Accept Them with a Smile

Whether you are a child studying in a chawl or a founder raising funds for your first startup, pain and rejections, in the early years, are the only recurring revenue you can reliably expect. Accept them with a smile on your face and fight on.

My loan application for education was rejected by more than twenty-one banks in Mumbai. For every bank we approached, the officers cared only about my parents' income, which, for many reasons, was far below the threshold required to secure education funds. My top-tier university ranking did not matter to them, nor did the admit from USC, because there is no checkbox for either in a loan application. Scores do not matter; capital does. Needless to say, my parents sold much of their assets and liquidated all of their retirement savings to fund my education abroad. With this, along with a few scholarships and the contributions made by a few blessed souls, I was able to pursue my education overseas.

In 2012, I landed on USC's campus with the hope to build a better life for my parents, my family, and for students around the world for whom education is but a luxury and dream. Once at USC, I quickly realized that, while my initial plan was to specialize in telecom systems, 4G was already launched in the US and 5G was decades away. This meant there probably wouldn't be enough opportunities for me to get a job post-graduation and repay my education debt. Going back home with a huge financial debt was not an option. So what do you do?

This is where I learned my sixth lesson on life and startups. When your original plan fails, learn to adapt, but don't lose sight of your target.

Lesson 6: Learn to Adapt & March Forward — Be a River, Not an Ocean

To save funds and increase the odds of developing a meaningful career, I made two changes to my approach. First, I decided to graduate in three semesters. Second, I decided I will diversify my course portfolio. I took pretty much one course from each domain, an approach that, to many, might seem like a career suicide. After all, what do you specialize in when you take one course from each domain? That's exactly the point. You don't specialize in one domain; you develop a semi-deep understanding of multiple domains. I took one course on computer networks, one on telecom, one on space communication, two on advanced Markov chains, one on fiber optics, and one that covered cloud systems.

As it turns out, this bet paid off big time in the years to come. If it wasn't for those courses, I wouldn't have been selected to design mission-critical systems for the governments of the US, UK, Norway, and the Netherlands. If it wasn't for the Markov chain course, I would have never understood modern AI and LLM models. The same applies to the prediction and path-planning algorithms we use in mobile robots, humanoids, and self-driving cars.

So, after graduating from USC, I took a job and spent the first two years building telecom systems for local US government agencies. After a year in this role, I had reached what some might refer to as a dead end in my career. My work was pretty much repetitive at this point, and it didn't take long to realize that this 10-12 hours' worth of daily non-stimulating exercise was diminishing the purpose of life and value of each day given to me.

I hadn't travelled from the chawls of Mumbai to LA — 10,000 miles — only to design telecom systems and program base stations.

That's when I decided to pivot to computer networks. This is where the seventh lesson comes in handy. If you ever want to fall in love, fall in love with the process of learning new subjects. Curiosity is a powerful tool, and when pointed in the right direction can open doors that no capital in the world ever will.

Lesson 7: Cultivate a Lifelong Love for Learning

Initially, I ordered a few thick books from eBay and studied those day in and day out — after work and before work. But over four months of doing this, I realized it felt more like reading an encyclopedia and less like a structured approach. So I changed my approach. I decided to take a more hands-on approach.

I went on eBay, ordered routers, switches, firewalls — and built a data center from scratch in my rented apartment in Los Angeles. First for myself, and then for governments.

It worked. The path wasn't easy, but it paid off.

After building this data center, I completed two of the most respected and expensive networking exams and was selected to design and implement a mission-critical data center and communication system for a government agency in the UK. Later, I did that for the governments of the US, the Netherlands, and Norway. From helping design and launch the equivalent of a private version of Google Maps, to a secure chat app, and a secure push-to-talk app in dedicated data-centers for government agencies, I had learned the A to Z of software applications, system design, and data-centers.

At this time, I wanted to expand my capabilities beyond telecom, networking, and software applications. Naturally, robotics and AI felt like the next move. How does one get into robotics and AI? Well, repeat the same process: build, learn, promote, and deliver deep value.

So I did the same. That brings us to our next lesson: always deliver deep value.

Lesson 8: Always Deliver Deep Value

I built a few robots after hours, starting from scratch. Self-taught Arduino, ROS, perception systems, Linux OS, sensor fusion mechanisms — the whole nine yards of robotics. Same for AI: ML, YOLO, neural networks. This was around 2018. It felt like I was drinking from a firehose.

Lo and behold, after I felt I had learned enough in a few months, I started strategically reaching out to robotics companies, showcasing my work and the desire to create value for them. I would draft one-pagers with problems they might have (as I understood them), and write a solution, then email and DM them.

Eventually, a team at an emerging Silicon Valley robotics startup funded by Bloomberg and Sequoia Capital interviewed me. After 6–7 rounds of interviews that covered subjects ranging from networking, operations, sales, program management, the philosophy of life, and debugging live problems in robotics in front of the founders — I did okay and was selected.

So I finally moved to the Valley in 2019, and that was the beginning of my chapter in Silicon Valley, robotics, and AI. The efforts that I had put into studying robotics after coming from work each day for months in a row had finally culminated in a role that gave me an opportunity to contribute to commercial projects in robotics. We will cover lessons from this journey in the upcoming chapters.

Before we get there, it's worth reflecting on how most people begin their journey in this field— by building some variant of a TurtleBot, a simple two-wheel-drive robot. Almost every founder and engineer working in this space can relate to this robot, yet few are aware of its history. Understanding the roots of this robot matters because it laid the foundation for all modern mobile robots, from the vacuum robots you see in homes to the delivery bots you see in cities. It's this robot that helped us build a solid architectural foundation for all modern robots we rely on today.

We will cover this story in the next chapter.

Chapter 02: Silicon Valley & the Evolution of Robots

Most engineers studying robotics tend to think that robots were invented in Silicon Valley. In reality, they were invented about 5,400 miles east of Silicon Valley, in Bristol, England. In this city, in 1949, neurophysiologist William Grey Walter built the first autonomous robots, i.e. Turtlebots — small, battery-powered machines that could follow a light source and steer toward it using sensors and a few lines of code.[1] It was a simple robot, but it navigated in a way that ignited the hearts of many robotics professionals around the world. Amongst these inspired souls, one of them was George Devol, who, in the 1950s, created *Unimate*, the first programmable industrial robotic arm that would go on to transform automotive factories in America.[2]

While Turtlebot and Unimate laid the foundation for commercial robots and helped automate a lot of physical work in the industry, they at the end of the day, had to be hardcoded. In simple words, they could follow rule-based instructions hardcoded on the onboard computer, but they couldn't *think* on their own. They lacked intelligence.

That changed in 1972, when engineers at the Stanford Research Institute built the first AI-powered robot called Shakey.[3] This mobile robot outperformed Unimate and Turtlebot. For beginners, this robot could sense its environment and develop a step-by-step navigation plan on its own. It was able to identify and move objects from one place to another on its own. It wasn't hardcoded, unlike Unimate and Turtlebot. It was trained to learn and improve with each task. Although it was developed at Stanford, the contributions came from all over the world, including Europe and Japan. From there, the thread stretched to Boston Dynamics' early work in the '80s and '90s, and to household names like the Roomba Vacuum Bot (2002), Kiva Systems' Warehouse Bot (2003, now acquired by Amazon), Starship's delivery bots (2014), and Cobalt's security robots (2016).

But if the idea of building intelligent robots like Shakey was global, the platform that would power the next wave of robots was indeed born in Silicon Valley.

In 2006, a small lab in Menlo Park called Willow Garage laid the foundation for modern robots by building ROS — the Robot Operating System.[4] As some of you might be familiar, it is equivalent to the Android OS for mobile phones, except this one is made for robots. Although ROS didn't solve all core problems in robotics, it gave builders a common language and a standardized framework to build modern robots. It ensured anyone with a laptop equipped with Linux OS could spin up a ROS node, i.e., code that can steer a two-wheel or a four-wheeled robot using a keyboard. It made some serious robotics projects possible outside of top-tier labs. It gave birth to the tsunami of robotics startups in Silicon Valley, each powered by roboticists intending to bring robots into everyday lives.

From Anki to Mayfield and Astro, these robotic projects serve as core reminders of lessons that, to date, govern the field of consumer robots. So, what are these lessons? We will start with Anki.

Building Robots with a Soul - Lessons from Cozmo and Vector

Anki was founded in San Francisco in 2010 with a bold mission — to create consumer robots that could emotionally engage with humans.[5] In simple terms, they were on a mission to add *soul* to robots so they could serve as trusted companions to both kids and adults.

This mission came to life through *Cozmo* and *Vector*, two small robots packed with human-like personalities. *Cozmo*, their first robot launched in 2016, was blessed with sophisticated features, including but not limited to the ability to recognize and remember faces, express intense emotions such as delight or disappointment, and learn from repeated interactions with kids. Beyond these interactive features, the team building Cozmo went a step further and developed a graphical drag-and-drop programming interface to make it as simple as possible for school kids to program this robot. It's this programming interface that empowered the team to add foundational AI modules in the next robot they launched in this series, i.e., Vector.

Released in 2018, *Vector* was Anki's more sophisticated "home robot," designed to be an autonomous, AI-powered companion. Unlike Cozmo, it had its own onboard processor, microphone array, and the ability to seamlessly interact with its environment. If you look at these robots — many of which are still available online — it's hard to fathom the fact that products so well designed and launched with such love and craftsmanship could still fail commercially.

While these robots were built for our homes and classrooms, to introduce families and children to the idea that a robot could be just as curious about everyday life and carry a sense of humor, just like us humans, they failed to make an impact. Vector and Cozmo succeeded in developing a strong human-robot interface channel but failed to create deep value for end users. They failed to solve a real problem.

The challenge wasn't the technology — it was the business model. A companion robot that is solely meant to be a companion is, to date, a product led by desire and not a need. If there is no need, no amount of desire can help you win the market. Beyond addressing this product-market fit problem, hardware margins were thin, recurring revenue was limited, and the excitement of owning a cute, interactive robot faded once the novelty wore off. Given these challenges, even after raising nearly $200 million, Anki — like many other consumer robotics startups — shut down in 2019.

Perhaps, in a way, it wouldn't be wrong to say that Anki laid the foundation for building robots with strong human–robot interfaces. But that novelty didn't translate into sustainable revenue.

Its story remains a quiet but powerful lesson in robotics: emotional design can win attention, but enduring success requires deep utility. Consumer robots, like industrial ones, can only succeed if they solve a real problem. Unfortunately, unlike phones or cars, consumer robots don't get as much leeway when it comes to adhering to the line that separates *need* from *desire.* Consumers are willing to pay a premium for a phone that costs as much as a laptop, even if an alternative, cheaper model is available with similar functions — but that dynamic doesn't exist in robotics. Unless consumer robots solve a real, and an extremely painful problem, long-term adoption will always remain a challenge.

That being said, building on this early wave of strengthening human–robot interaction and creating robots that could coexist with humans in daily life, another company tried a similar route.

Elevating Robots with a Soul - Lessons from Mayfield Robotics

Mayfield Robotics, a startup backed by Bosch, shared a vision similar to Anki's.[6] It started with a goal to bring robots into our homes and daily lives. The company's mission was built on the belief that emotional connection, combined with elegant design, could redefine how humans relate to robots. Their creation, *Kuri*, a companion robot, embodied that vision.

Launched as a consumer home robot, Kuri looked like a perfect Pixar character on wheels. The design, color palette, expressive eyes, and overall form factor were in line with what many of you as engineers might aim for in a consumer robot. Mayfield was betting that emotional presence and interactive capabilities would be enough to accelerate adoption of this extraordinarily designed product. While those two elements are core ingredients for helping a consumer robot attract strong attention in the market, without clear utility it's hard to translate that attention into lasting adoption.

The challenge here was that, just like we learned from the story of Anki, it's hard to replace utility with elegant design and interactive capabilities. For consumers, the practical question remained: *What does Kuri do for me every day? What problem does Kuri solve for me?* Beyond companionship, there was no clear function that justified her cost or permanence in a home. The emotional appeal that made Kuri special also became her constraint. She was delightful, but not essential.

By 2018, Mayfield Robotics ended operations.[6] The decision wasn't a failure of design or a lack of ambition; it was a recognition that emotional resonance alone cannot sustain a hardware-intensive business. Kuri's story, like Anki's, underscored a pivotal lesson for the robotics industry — that emotional design must be coupled with deep utility. Continuing this momentum, another consumer robotics project became a victim of this philosophy.

Lessons from Amazon's Project Astro & Fetch Robotics

Even Amazon, with all its muscle, struggled to gain adoption for its consumer robot. In 2021, it launched Astro — a mobile assistant, i.e., a consumer robot that looked like it rolled out of a Pixar movie. But Astro never became a household name. It worked, technically. But again, the same question loomed: *why does this need to exist? Why do we need a security robot moving on wheels scanning our home? Why not just put cameras?* Project Astro was shut down by Amazon within a few years of its launch.[7]

While most consumer robotics startups failed to take off during this time, the ones targeting warehouses and factories were quietly succeeding.

Founded in 2014, Fetch Robotics focused on warehouse automation.[8] Fetch didn't care about personality. It cared about payloads. It built autonomous mobile robots that autonomously transported packages in warehouses. The market was clear. The ROI was evident. Fetch was eventually acquired by Zebra Technologies.[8] It was one of Silicon Valley's rare robotics wins.

This acquisition made it clear that, for robotics companies to succeed, they must start with a real problem that customers are actively struggling with and urgently need to be solved. Building consumer-focused robots without validating the problem first is a fast path to failure.

Given these lessons, companies such as Figure AI, Tesla, Neo, and the Toyota Research Institute are developing robots that can automate factory work, fold laundry, and assist ageing adults. These are problems that are hard to solve and yet demand urgent and scalable solutions. Why?

Ageing population, shrinking skilled workforce, and changing demographics around the world have resulted in labor shortages that directly impact factories, warehouses, and healthcare service providers. We need more trained staff in these areas. Yet we are way behind, as we don't have enough people interested in doing these jobs. That's fair. It's just not fair if we don't solve this problem.

Given these lessons and areas that demand robotic solutions, one such company in Silicon Valley was focused on building safety and security robots. The value proposition was clear — we don't have enough security professionals to fill up the roles, and we need a way to augment this workforce with better capabilities. Cameras are good but don't provide any interactive or deterrent capabilities. Given this need, a safety and security robot might just address this problem. This was the vision of the company in the Valley.

After having interviewed around 2018, with the founders here, I decided to join this team and kickstart my professional career in this domain in Silicon Valley later that year. These years were filled with lessons that I have carried with me since then, and I hope to share those in the coming chapters one by one.

Chapter 03: The Art of Building Robots in Silicon Valley

The Valley is known for its upbeat optimism and endurance. You can take a stroll down El Camino Real in San Mateo and, almost effortlessly, bump into a handful of robotics and AI founders chasing their next big breakthrough. My path crossed with one such inspiring duo in the fall of 2018.

By then, I had already studied the foundations of robotics, built a few Arduino bots, and spent late nights researching Amazon's $775 million acquisition of Kiva Robotics and the evolution of iRobot's Roomba, i.e., the vacuum robot.[9] That curiosity soon turned into a mission. I began reaching out to robotics startups with a one-pager describing my understanding of their product, their bottlenecks and how I could help them unblock those. As they say, if you're willing to chew glass, doors will open for you; but if you're content chewing grass, even the strongest key won't open a single one. Guided by that belief—and perhaps a bit of monastic devotion—I reached out to every founder I could find in this space.

One such team stood out for their approach and philosophy. From the outset, my conversation with them transcended the soul-degrading discussions around revenue, funding, and tech that are common during such interviews. I interviewed with these founders of a small robotics startup in Silicon Valley and was fascinated by their vision – to make this world a safer and secure place.

What intrigued me most was their genuine interest in understanding my philosophy on life. That was, in fact, one of the questions I received from one of the co-founders of this startup during my interview. I remember my reply: "It's all about health, family, and career — in that order. If there is no health, what's the point of starting a family? And if there is no health or family, a career is a meaningless pursuit."

After a long discussion with each of the members of this startup team that continued for about a total of six hours, I was both fascinated by their vision and their philosophy. Subsequently, I decided to join this startup, Cobalt Robotics. Our project focused on developing indoor robots designed to automate routine tasks while providing safety and security services to people, properties, and places through cutting-edge robotics and AI systems. These robots are capable of patrolling office spaces and warehouses, complementing existing safety measures by providing real-time situational awareness. At the time, these robots were equipped with advanced sensors like LIDAR, 3D vision cameras, ultrasonic sensors, infrared cameras, and more, making them fully autonomous.

As exciting this venture might sound to most of us, it wasn't without its challenges. We are talking about a product that demands hardware, software, autonomous mobility, and continuous access to the internet, all with mission-critical robustness. A product like this is extremely hard to build, launch, operate, and, most importantly scale!

My job was threefold — ensure our product doesn't break in the middle of demos; ensure our fleet of robots deployed across the world is fully active 24×7; and ensure each new launch that we do is successful — no major incidents or failures, at least in the first two weeks post-launch and ideally never. We will now cover each of these scenarios one by one.

Demo days are exciting and challenging at the same time, not because you would need to drive your robots through the tech arteries of Silicon Valley, Texas, and Massachusetts, but because each demo meant an opportunity to create value for our prospect, investor and employees and last but not the least, an opportunity to learn how to not fail again. I encountered a decent amount of failures early on — from a robot dying in the middle of a demo due to indoor internet blind spots, a last-minute software release that an engineer accidentally pushed to a robot that was actively being demoed, a sharp ray of sunlight that hit the sensors of the robot so hard that they wouldn't function as intended — basically all sorts of challenges you can imagine from a mobile 120 lbs. device that's intended to function autonomously in an indoor space. Over the period of next few months, each of these lessons was iteratively shared internally with the team and woven into our product operations protocol.

Then came the post-sale launch periods. These are thrilling yet daunting times when a client decides to trial our service for six months or a year before fully committing. Think of demos as short sprints, while these phases are the long marathons that continue for at least six months and, in some cases, for years. Now imagine the challenges we faced on a demo day, but extended over 180 days, with the expectation to hit every performance mark. This phase is crucial for proving our worth and maximizing value for our clients.

Getting clients to test a beta version of our service was a tough task, making early adopters key to our success. This meant being agile enough as a startup to tailor the service to the needs of the client, even if it involved training a robot to navigate safely through an elevator, working with a telecom vendor in Australia, Japan, or the UK to get your robot to work on global telecom systems, finding the right shipping partner who could take care of customs and shipment, training people to fly overnight across the world to demo and deploy your service, and, last but not least, ensuring the robot's navigation mechanism is fail-proof and safe.

Yes, one of the hardest parts of building autonomous robots is to ensure they don't run into people. This one thing is easier said than done because, unfortunately, all the sensors that we use today in robotics have certain limitations and failure scenarios. From sun rays to extreme temperature, and viewing angles to the surface of the ground and its slope, and more, there are a lot of variables in this risk management equation that can become extremely hard to optimize for safety and security, with a focus on delivering real-time services.

Knowing this, the team from the get-go, introduced the human-in-the-loop concept. The goal here is that, while the robot will navigate autonomously, but for scenarios which we know it may not be able to handle with extreme caution, the robot would raise an alert and ask its navigator to help it maneuver. In doing so, the founders and the team ensured we delivered a service that was safe, reliable, secure and effective.

While sensor limitations, mission-critical robustness, and the scalability of a product that is both hardware and software-heavy were some of the challenges we encountered on the technological front, there were others that were more humane. We, as humans, aren't accustomed to interacting with robots, so it is completely natural to not have a taste for them. For instance, if you are a Fortune 100 company with 1,000 employees working on your campus, and if we introduced a safety and security robot to help your employees, not all employees would be willing to opt for such a service, and chances are most might either mock the concept or would be skeptical of such a service.

Thus, in robotics, while many think technology and operational barriers are the biggest challenges, I would argue that psychological barriers are probably the hardest to overcome. We all know we would much rather have a robot equipped with a deterrent tool and video-chat functionality to negotiate with an armed man than risk the life of a security guard, if we had the option, as our primary mode of resolution. However, the reality is that we humans have yet to develop trust and a taste for robots, and that's largely due to our evolutionary cycle and, partly, Hollywood — sorry, *Terminator* and *Transformer* fans. It reflects a broader challenge of familiarizing society with the real benefits of adopting a robotic assistant.

Through our robotic venture, we learned that these barriers can be navigated early on by conducting a series of pre-launch lunch-and-learn sessions with clients and employees, being transparent about how the robot works and why it does what it does, demonstrating multiple high-risk scenarios in a controlled manner, and sharing more case studies of clients who have benefited from these services immensely. Setting the right expectations and being transparent about what works and what doesn't with all your stakeholders is key for building trust and scaling your startup.

If you or someone you know is contemplating a career in robotics or plans to start a robotic startup, remember, robotic service is effective as long as you know exactly how it works, you have studied 99% of all failure scenarios, and have guardrails in place for the remaining 1%, and last but not least, you are transparent about the successes and failures of your service with your investors, clients, and employees.

That's what the founders did at this startup, and they are some of the most inspirational people I have met to date. While this chapter gives a bird's eye-view of what it takes to build a robotic startup in Silicon Valley, this journey is filled with galactic death traps— traps that can haunt even the most well-funded and talented teams in the Valley. We will now cover these lessons in detail in the coming chapters.

Chapter 04: Robotics Startups — Lessons You Learn the Hard Way

Building a robotics company is probably as close as it gets to chewing glass in the technology domain. Why? Basically, when you think about building a professional-grade robot, you are talking about building a human body using solely hardware and software, no biological component. You have to build the eyes and the ears, or the perception system; the legs or the locomotion and navigation system; the hands or the spatial manipulation system; and, most importantly the brain — the system that takes inputs from all the hardware components and makes decisions. Where should one even start?

The human body has evolved over years of evolution. We are blessed with extraordinary senses. Take eyes, for example, those two biological cameras you have been equipped with that come with specs of 576 MP each.[10] That's the power of the human eye. The same applies to the brain. When we refer to the brain, we are referring to around 100 trillion neural connections that allow us to connect the dots in our life and make decisions every second. That's your thought pattern, powered by 86 billion neurons—each forming thousands of connections, resulting in a total of about 100 trillion.[11]

Now, given the power of our human biological system, it's no surprise that when you try to build one using sensors, hardware stack, and software, it's unlikely that you will develop capabilities that are even 10% close to the human body. More importantly, these robots need to operate among us humans. There is no doubt they are going to make mistakes, just like a toddler, and some mistakes can be fixed while others can result in a death trap for even the most promising robotics startup.

While a lot is being done to accelerate the adoption of robots across different facets of our lives, there are known risks in this space that frequently catapult promising robotics startups into resource-constrained mode, eventually shattering the dreams of both team members and founders. It's an emotional journey, especially when you know you have the right product, but you have no way to proactively identify these risks, i.e., traps, before they do considerable damage and exhaust all your resources.

To better understand these traps, it helps to zoom out and understand the journey of robotics startups. If you study the journey of all major robotics startups, they all go through these core phases:

- **Early adoption and growth phase**—where everything feels possible.
- **The accelerated resource drain phase**—rapid loss of capital, time, and talent
- **The survival phase**—where, despite the best efforts of teams, each day feels like an uphill battle against time. This is where you will learn a ton of lessons that will most likely shape the trajectory of your robotics startup.

We experienced these three phases and learned the corresponding lessons that came with them the hard way a few years ago in the Valley. We had basically designed a sophisticated version of TurtleBot to automate safety and security services in offices and warehouses for startups and Fortune 500 clients. Our 5 ft+, 120 lb. robot was capable of autonomously patrolling these spaces and was aimed at augmenting safety and security staff. To get to the point, we were trying to help companies that couldn't hire reliable safety and security teams to add robots so they would need fewer team members and could do more with robots. These robots were equipped with advanced sensors like LiDAR, 3D vision cameras, ultrasonic sensors, infrared cameras, and more, making them fully autonomous.

A product like this is extremely hard to build, launch, operate, and, most importantly, scale. My job was threefold: ensure our product doesn't break in the middle of demos; ensure our fleet of robots deployed across the world is fully active 24×7; and ensure each new launch that we do is successful, with no major incidents or failures, at least in the first two weeks post-launch and ideally never.

During this time, I realised running a robotics startup is brutally hard. Every mistake is costly, and every lesson comes at a significant price. We made plenty of mistakes — some avoidable, some inevitable. This chapter is about the lessons we learned that I hope will help future robotics founders, engineers, and investors save a ton of time, health, and capital. I hope it helps you navigate the toughest parts of this journey with fewer scars. We will now cover each of these lessons one by one.

1. The Long-Tail Problem in Robotics Is Unforgiving

No matter how much you optimize your sensors, perception and navigation system, you won't catch everything. The long-tail scenarios, i.e., edge cases in robotics, are unforgivingly exponential.

Take LiDAR for example—a $10,000 sensor that's often seen as the gold standard in robotics. Except when you realize it is not. We learned this the hard way. Each of our robots was equipped with a LiDAR that came with a hefty price tag of $10,000 at the time, and yet all it took was a mere stream of photons, i.e., light particles emitted by the sun, to make it blind. This happened during a pilot run when one of our robots, navigating near a staircase, was completely blinded by sunlight pouring in from a nearby window. In this state, it is probably as dangerous as a car without brakes.

There's a solution to this, but it's not perfect. You can add no-drive zones to your robot's map and ensure a wide buffer area between driveable and restricted areas. It's a workaround, but in robotics—especially in a startup—workarounds are often your best bet. This isn't the only workaround you'll need. If your robot keeps losing its position, you have a bigger problem–Localization.

2. Localization: Easier Said Than Done

ROS makes localization seem easy, until you realize it is not. This means in the early stages of real-world deployments, your robot will frequently cross into no-drive zones, bump into walls, people, or other obstacles.

For startups building indoor robots, localization failures are inevitable, and when they happen, robots will drift into places they shouldn't. We saw this happen in warehouses and office spaces with long, tunnel-like hallways. When mapped in ROS, these tunnels appear as 2D rectangular spaces on the navigation map. A robot moving continuously in a straight line on such a map will eventually lose track of its position, no matter how good the navigation system is.

Our solution? Strategically placed QR codes along hallways to help the robot recalibrate its position every time it encountered a QR code. It worked, but it required constant calibration of sensors and extensive site-specific tuning. That's the reality of robotics: solutions exist, but they come at a hefty cost. If you are not careful, you can easily drain your resources in this endless loop of sensor calibrations and site-specific tuning.

And that leads us to the next lesson—the customization trap.

3. Scaling Smart: Fewer Clients, More Efficiency

In the first two phases of your journey, initial buildup and growth, resist the temptation to take on too many clients. It's a trap.

With every new client comes a flood of edge cases. This is especially true in robotics. For every one you solve, a hundred more will surface. Soon, you're developing custom solutions for every site, every map, every hallway, office space, home, staircase, elevator, and escalator. This is the single biggest drain on your resources—because custom solutions don't scale.

Instead, go deep, not wide. Prioritize clients with cookie-cutter operational zones—consistent physical environments where your robots will operate. Work with a few clients who own multiple, similar properties. Think Costco warehouses, Walmart stores (after-hours cleanup), or Target (robotic-cart-enabled shopping and checkout). These environments don't change much.

Scaling within such cookie-cutter spaces slashes customization costs and eliminates 90% of the operational overhead that drains most robotics startups. If you get this right, you can scale faster with far fewer headaches.

There is one more reason why this approach is startup-friendly. If you are in the B2B robotics space, a vast majority of your clients will demand integration with on-site devices and sensors. And this is much easier done if you have a few clients with a consistent setup across their sites, so the fix you develop for one can be deployed at others. This means making your robot autonomous is just step one; integration is what matters, and that leads us to our next lesson.

4. Autonomy Is Just Step One—Integration Is What Matters

For startups, providing robotics-enabled services in indoor environments like offices and warehouses, autonomy alone isn't enough.

Your software must integrate with the client's ecosystem. No exceptions. For example, in our case, during our early days, very few clients were willing to accept a system that didn't connect with any of these: their elevators, safety systems, camera networks, and card readers. Eventually, we had multiple integrations in place. We realised that integration is one of the core client needs that you will need to address, especially if you are a robotics-enabled service provider operating in the B2B space.

If you plan for this early, you save capital and engineering effort. If you don't, you'll be scrambling to retrofit these features later at 10× the cost. Speaking of the 10× formula, for those of you that are running or have led a robotic startup, you would probably agree that running a robotics startup indeed feels like running 10 companies, and that leads to our next lesson.

5. Running a Robotics Startup Is Like Running Multiple Companies

As our founders often put it, running a robotics startup is like running multiple companies at once—hardware, software, networking, AI, and more.

Let's start with hardware. Wherever there's hardware—computers, sensors, batteries, chargers—failures are inevitable. And when you're operating fleets across the country or internationally, repairs and outages become one of your biggest operational expenses. You won't always have the financial resources to deploy people to every site to fix every robot in time.

The only way to survive? Build redundancies into every layer of your tech stack.

If local Wi-Fi goes down, you still need a way to track the robot's heartbeat and ensure safe navigation. If a sensor malfunctions, the system should auto-compensate without shutting down operations. That's why we used LiDAR, RADAR, ultrasonics, and cameras to navigate obstacles, but even that isn't foolproof. Each sensor comes with its own blindspot, and the union of all doesn't guarantee safe navigation at all. That's why a human-in-the-loop is essential in robotics, a concept we covered in the previous chapter. That human operator will serve as a backup in case your robot's onboard algorithms malfunction. The same concept of redundancy applies to the network connectivity module of your robot.

If a local Wi-Fi network goes down, you still need a way to track the robot's heartbeat and ensure safe navigation. This means developing software and hardware to connect your robot to telecom networks so they can serve as a backup source of connectivity when Wi-Fi fails.

This is easier said than done and often turns into a massive project on its own, especially for robotics companies that intend to grow their operations across different geographies, countries, and continents. If you are in this boat, you may want to rethink your strategy. This is where the next lesson comes in handy.

6. Global Expansion in Robotics: The Early-Stage Trap

At our startup, we expanded our robotics services across the US, UK, Australia and a few other countries. This strategy might sound exciting in theory, but when executed in reality, leads to an operational maze that can easily become a nightmare. You're dealing with delays in shipments, variations in telecom network standards, country-specific regulations, and an endless list of logistical challenges. What seems like an exciting milestone often turns into a slow, costly distraction from your core execution path.

Before scaling across borders, ask yourself: Have you truly nailed product-market fit in your home country? Do you have a repeatable, scalable model? If not, expanding too early won't accelerate growth—it will amplify risks and inefficiencies. While there are many of these risks that we can explore here, in the interest of time we will focus on a core one that arises with the global expansion: robot connectivity.

Getting robots to connect to telecom networks means working closely with telecom vendors, procuring hardware modules, SIM cards, and understanding how local operators have configured their networks. Integrating these details into the robot's operating system remotely, either before shipping the robot abroad or on arrival, can be extremely challenging.

I led this effort for 2.5 years and quickly realised that connecting robots to local telecom networks is just the beginning. The real challenge is ensuring that every layer of networking software — encryption, time-outs, feed prioritisation, network selection, and more — remains synchronised and functions reliably across different telecom networks and countries.

Similarly, hardware modules (robot modems), firmware, and telecom operators vary significantly across countries. This means you will now need to meticulously track network hardware and software configurations for each robot in your fleet for each geographic region.

And it doesn't stop there. Even within the same country, network performance can be highly inconsistent. In the U.S., for instance, T-Mobile may have stronger coverage in some cities, while Verizon or AT&T perform better in others. Expanding globally means adapting to these operational variations. You will need to plan ahead, or connectivity issues will quickly become your bottleneck.

Speaking of bottlenecks, technical bottlenecks are generally easier to address — if you have the right team structure in place. But building that structure is easier said than done.

7. The Hardest Part? Building the Right Team

Running a robotics startup isn't just about having the right people; it's about having the right team structure. Each team member should have their strengths, but they also need to be capable of handling tasks outside their core expertise.

For example, if you have a robotics engineer specialising in ROS — navigation, perception, localisation — it wouldn't hurt to give them a basic 101 on network debugging. Similarly, if you have a network engineer handling all things connectivity, they should learn enough ROS to troubleshoot when needed.

In my role, while I wasn't expected to code much, I built enough hands-on skills where it mattered. I could debug ROS nodes, debug navigation issues, troubleshoot network connections, write custom networking scripts, deploy the latest code to our robots, and debug hardware and software issues through the terminal. Not because I was expected to, but because time is everything in robotics, and relying solely on experts in a crisis will not save the day — certainly not if you are the one in charge of leading demos and deployments for clients.

Working closely with our clients, managing product demos and deployments, I knew from day one that I had to understand all major components well enough to debug issues myself, at least as much as possible.

That's because, if you are running a robotics startup, you understand that time is a scarce resource. Every demo is a hedge against the failure of one of thousands of components and software services that make your robot valuable to a client. Failures are inevitable, but learning from them early, feeding those insights to the right teams, and fixing underlying software issues is a far better use of time than escalating every single problem for an immediate fix.

In our case, this approach kept engineering teams focused on long-term improvements instead of constantly putting out fires. Being mindful of everyone's time is one of the most valuable lessons a team can learn. This approach keeps teams focused, cross-functionally trained, and ensures that when things go wrong, your startup isn't paralyzed because a specific expert isn't available.

Cross-functional teams also help mitigate talent drain, one of the biggest challenges in robotics. The field is still evolving, and companies go through cycles of rapid growth and sudden downturns. This means building a team that is flexible, resilient, and capable across disciplines is the key to long-term survival.

Robotics startups are brutal — for both founders and team members. But if you learn fast, adapt quickly, and build a resilient team, you give yourself a real shot at survival.

Speaking of survival, now that we have covered some of the core bottlenecks that most robotics startups face, it goes without saying that the hardest part of this journey is landing the right client. This is probably one of the most brutal processes in this journey, as one mistake here can easily drain your capital lifelines in a matter of months and bring you close to a full-blown shutdown.

So how should a robotics startup or an engineer identify and approach clients? We will cover that in the next few chapters.

Chapter 05: When Apple Said No to Our Robot

A lot of startups tend to pursue Fortune 500 customers way too early in their journey. It seems like the right thing to do—big names, big contracts, credibility—but the reality is a bit different than what you would think. If you're not careful, you can waste months, sometimes years, without closing a deal.

The focus of this chapter here is specifically on dealing with different types of customers without draining your resources. We'll keep it simple: prospects fall into two buckets—Fortune 500 Giants and fast-moving mid-sized companies.

1. **Fortune 500 Giants** – Big logos, long sales cycles, and a ton of stakeholders. Winning them gives you credibility at scale, but it also demands extreme endurance and patience.
2. **Mid-Sized Heroes** – Fast-moving companies that offer speed and agility. They're quick to decide, willing to experiment, and often form the backbone of early growth for startups trying to make their mark in the market.

How you handle these two from day one, whether it's a demo, a sales pitch, or six months of chasing your seventh meeting with the same prospect, will determine your startup's survival.

In this chapter, we will focus specifically on lessons we learned through our meeting with a Fortune 500 giant, i.e., Apple.

Before our meeting with Apple

Oct 16 and 17, 2019, and the months preceding this meeting with Apple, were crucial in our journey as a startup. They taught us some hard lessons around evaluating prospects, understanding their real needs early on in the process, and being deliberate about where we invest our time and resources.

At the time, we had a few paying clients, mostly from the technology space, but our product had yet to find a solid product-market fit.

Our product worked. It did exactly what it was designed to do: autonomously patrol office spaces, warehouses, and other facilities while delivering real-time safety and security insights to security and executive teams.

The challenge?

Consistency within a single industry vertical. We needed a steady stream of clients in one core domain—healthcare, technology, warehousing, or pharmaceuticals. Instead, our clients were scattered across a wide range of industry verticals: a food delivery startup, a financial company, an e-commerce brand, a technology firm, and more.

This was both good and bad.

Good, because it validated demand across multiple industries.

Bad, because it meant an endless cycle of customizations and diverse standard operating protocols (SOPs) for each client, straining our engineering, operations, and sales teams, and significantly throttling our ability to scale.

It was during this time, fall of 2019, that our paths crossed with Apple.

The Apple Meeting: How It Happened

So around this time, in the fall of 2019, after months of persistence navigating a long chain of internal champions and a maze of intros, one of our sales team members secured an in-person meeting with the decision-makers at Apple headquarters. As someone responsible for leading the demo, I had a sense of what was at stake here. Success meant objective validation from a high-caliber Fortune 500 giant. Failure meant valuable lessons, but at the extreme cost of capital, time, and effort.

That reality drove a series of risk-focused exercises: rigorous product testing, strategic briefings with engineering, sales, and operations teams, and the intensified adrenaline rush that comes with a high-stakes deal.

We spent a reasonable amount of time preparing for this meeting. It took place on 17 October 2019 at Apple HQ in Silicon Valley.

To be clear, all this while, despite our best efforts, we were only able to speak a handful of times with below-the-line contact, someone close to the decision-maker but not the final authority. They seemed deeply interested in our service and helped us secure this meeting.

Over the months leading up to this point, we believed they were looking to augment and automate their existing security systems with robots, and eventually this contact got us in front of the real decision-maker.

We had a one-shot.

We were ready—or at least, that's what we thought.

On October 16, the day before the big meeting, we arrived at Apple's Main Campus. We met our internal champion, reviewed their needs—at least, what we thought were their needs. Then we spent hours making sure we were fully prepared for the meeting: product testing, revisiting the problem statement, rehearsing our approach, and covering every detail we possibly could.

Then came the big day—October 17.

On the day of meeting–October 17

On the day of the meeting, we rolled the robot into Apple's office, brought it to the meeting room, and greeted the five or six people from their team who had joined. The meeting began, and we reiterated our understanding of the problem they were trying to solve before moving into the demo.

We then explained how we could help them address those challenges and demonstrated our robotics-enabled security service to their team. The robot performed flawlessly. The video-calling service, the badge identification system, and the safety and security anomaly detection system all operated without a hitch. By this point, we had built enough trust to show them that our product had the potential to address their needs and deliver long-term value.

Then came the deal-breaker.

The Deal-Breaker

There are days when you know exactly where your product stands—its strengths and weaknesses, the things it does exceptionally well and the things it still struggles with. We knew that we had one of the best autonomous navigation systems available, meticulously tuned for precision. Our hardware was resilient, with sensor redundancy at all levels. Our security compliance was solid, and our anomaly detection system was fine-tuned with some of the best machine learning models available. Even the creator of the YOLO AI model might have appreciated how precisely we fine-tuned it for object detection in the context of safety and security service.

But there was one thing we knew we couldn't do, and that one thing defined the outcome of this meeting.

Where It Fell Apart

At this point in the conversation, one of the Apple executives asked:

"How many floors can one of these robots cover?"

Apple didn't want one robot per floor. They wanted one robot per building, capable of navigating multiple floors autonomously. That meant integrating every single robot with elevators.

Placing one robot on each floor was not an option since that would mean transforming a campus into a dystopian city with 100+ robots per campus, one per floor. The robot had to handle elevators.

Our reality?

At the time, our robots weren't integrated with elevators, certainly not with the ones Apple had installed. Our 2WD, 5 ft 4 in, 120 lb. robot couldn't climb stairs or use escalators either.

This was not a minor feature request. This was a deal-breaker for Apple.

We had two choices.

The first was to commit to building elevator integration, fully aware that it would take months, drain resources, and delay existing deployments. The second was to be honest that we couldn't deliver within the required timeframe and walk away from the deal.

The decision was clear. We had to walk away.

This is how startups die. You think you understand what a customer needs, only to realise—often too late—that you don't. Enterprise sales cycles are brutally slow, and the process can burn months of your most valuable resources, especially if you are an early-stage startup: your time, your team, and your capital.

Even knowing this, when you're in the arena, the temptation to engage with industry giants is strong. We felt it too during this engagement.

The Hardest Lesson in Selling to Fortune 500s

Walking out of Apple's headquarters (HQ) that day, as I was driving back to our startup HQ with a robot in a big van, it was painful to digest this outcome. Not because it led to a failure, but because of the amount of time and effort we had put into preparing for this one meeting.

It was an important one for us, and yet it hadn't worked out.

The failure was not in our product; it worked flawlessly. It was not in our pitch; we built trust and engaged the room. The failure lay in not investing enough time to understand the real needs of our prospect with an extremely high degree of objectivity and clarity.

This was a hard blow for me and my team.

If we had been able to make this work, it would have changed the trajectory of our company. Our product could have been deployed across campuses nationwide. But in hindsight, the real lesson was not about the technology; it was about understanding what your customers truly need before you invest time and resources chasing them.

We spent months navigating Apple's internal decision-makers, preparing for every possible objection, and fine-tuning our demo. But in the end, we weren't solving the problem that mattered most to Apple.

Had we known from the outset that Apple's non-negotiable requirement was multi-floor navigation with elevator integration, we wouldn't have wasted time chasing a deal we could never win.

While there are many lessons we can learn from this engagement, and probably more insights into this event, at the risk of not boring you with all the details, here are some of the most valuable lessons we learned from this engagement.

1) Landing a Fortune 500 Meeting Is Easier Than You Think

Many founders believe that securing a meeting with a Fortune 500 company is the hardest part. That might be true in some cases, but the reality is that if you understand the core problem of your client with extreme depth and clarity, landing a meeting with Apple, Amazon, Toyota is not that hard. After all, who doesn't want a solution that genuinely solves their problem? If your product is compelling and solves a high-stakes problem, decision-makers will listen and make time for you.

The real challenge begins once you're in the room. What matters is how well you understand the client's world before you step into that conference room: their constraints, priorities, workflows, and deal-breakers. Which brings us to the next lesson: committing yourself to deeply understanding the needs of your client.

2) What They Say They Want ≠ What They Actually Need

Enterprise clients operate with layers of priorities, budget profiles, and internal constraints, which means what they say they want and what they actually need are rarely the same. We spent months chasing Apple, yet despite our best efforts, we never got on a call with key decision-makers before the demo to truly understand their core needs.

We assumed Apple wanted a robotic-enabled security system. In reality, they wanted a scalable one—specifically, one robot per building, not one per floor. That meant elevator integration was non-negotiable.

This realization came too late. We didn't ask the right questions when it mattered, and we paid the price. The real lesson here is simple but unforgiving: unless you understand the client's must-have requirement at the very beginning, you will waste time, talent, and capital pursuing a deal that was never in your strike zone. Clarity early on saves months later.

3) Fortune 500s Don't Buy the Best Product — They Buy the Best Fit

A great demo doesn't guarantee a sale. We demonstrated how our autonomous robots could help amplify safety and security services across Apple's campus. Our robot worked flawlessly during the demo. But Fortune 500s don't buy because a product works; they buy because it fits into their existing world. It must be compatible with their needs and their infrastructure.

Apple wasn't evaluating our tech. They were evaluating whether we fit into their infrastructure. We didn't. Enterprise adoption boils down to compatibility. If your solution doesn't integrate seamlessly into their existing environment, or if it introduces operational friction, the deal is dead before the conversation begins. Fit outweighs features every single time.

Beyond compatibility, one of the most important lessons to learn here is that enterprise deals have low conversion rates because technology is only one variable in this sales equation. Internal alignment, budget cycles, vendor-risk assessments, and IT security reviews matter far more than most startups realise.

4) Enterprise Meetings Aren't Discovery Sessions — They're Exams

Meeting with Fortune 500 executives is brutally time-constrained. During these meetings, you are not just demonstrating your product or your service; you are fighting for attention in a room full of high-stakes decision-makers with packed calendars. If your insights do not immediately resonate, you are already forgotten.

These meetings are not meant to be an extended discovery session. You need to walk in with a level of understanding that makes them feel as though you are already an insider. They want to see that you have done your homework, that you have studied their pain points and infrastructure constraints with extreme clarity before you step into the demo room.

That means:

- Understanding their infrastructure constraints before they tell you
- Understanding how your solution can be integrated into their existing workflows
- Anticipating and addressing deal-breakers before they bring them up

Mid-market clients will give you room to figure things out; you can iterate with them, adjust, and refine your offering as you go. Fortune 500s will not offer that pathway in most cases. Hence, the worst time to discover a deal-breaker is during these meetings.

By the time Apple asked about elevator integration, the deal was already dead. These meetings with top-level decision-makers are a litmus test to see how well you understand their world. If you have to ask during the meeting, you have already failed. No exceptions here. This was one of the hardest lessons we learned.

5) Be Strategic About Your Sales Pipeline – A Slow 'No' Is More Dangerous Than a Fast One

You really want to be strategic about your sales pipeline because that will define the type of clients your team will chase. Chasing enterprise clients is a high-risk, high-reward game. It often comes with a slow "no", and a slow no is more dangerous than a fast one. Why? Because you have a limited supply of talent, capital, and, most importantly, time. If more than 20 per cent of your sales pipeline is dependent on Fortune 500 clients, you have a big problem.

Define upfront what percentage of your revenue should come from Fortune 500s versus mid-market clients. If too much is riding on Fortune 500s, your startup becomes hostage to slow-moving deals—deals that could stretch months or even years without a clear outcome.

Startups move fast. Fortune 500s do not. We spent months getting in the door, then months prepping for the demo, only to realise, too late, that this deal was never viable in the first place. Had we prioritised mid-sized, fast-moving, high-fit clients in our pipeline, we could have saved time, closed more deals, and thrived.

6) The Real Lesson? Ask the Right Questions Early.

Our biggest mistake wasn't losing the deal; it was thinking we knew what Apple needed without asking the hard questions early.

Want to avoid this trap? Ask at least these three questions before you spend a single minute on an enterprise deal:

1. What's their absolute must-have requirement?
2. Does your product in its current state meet those requirements?
3. If not, how much time and effort would it take to get there?

If the answers aren't clear, steer away from that deal and take time to reflect on the deal-breakers both for you and your prospects. Always remember that a slow "no" drains more resources than a fast one.

Not every Fortune 500 deal ends like this. In fact, a different enterprise customer, a top US pharmaceutical company, turned out to be one of our biggest wins. More on that in the next chapter.

Chapter 06: Why Enterprise Deals Live or Die on Security

Working with enterprise clients as an early-stage startup is a high-risk, high-reward game. Success depends on deeply understanding your prospect's must-have needs. When you are working on such deals, there are times you learn from failures, and there are times when your dedication and persistence pay off. That is exactly what happened in our engagement with a top pharmaceutical client.

In this chapter, we'll cover this engagement and the five core lessons that came with it:

- **Why security is not optional;** a strong security posture is the first filter for enterprise clients.
- **You don't need a full-blown security team:** we partnered with a security audit firm, and they delivered deep value.
- **Your hardware tells a story:** make sure it is the right one.
- **When to adapt and when to push back;** not every enterprise demand is worth accommodating.
- **How to make these calls in real time;** navigating high-stakes decisions when you are in the arena.

So, how did we secure this deal, and how did we execute it? We will now cover this journey step by step.

Meeting with a Pharma CIO

Working with Apple helped us understand the importance of deeply understanding a prospect's core needs and the consequences of not doing so. We covered this engagement in the previous chapter.

Later, in 2020, we crossed paths with a top-tier pharmaceutical client—one of the world's top five, based in Billerica, Boston, US. This engagement became a valuable startup lesson for us. It helped us understand, for the first-time, the sheer amount of effort it takes to earn the trust of a cautious CIO and turn them into a strong advocate of your product.

It was during this time that we learned that credibility in the enterprise space boils down to three things: a rock-solid security posture, the ability to adapt, and the judgment to know when to adapt and when not to. For those curious about how we navigated this storm, it's worth understanding the context here.

The Setup: A High-Stakes Enterprise Demo

In 2019, I was running an important demo for the Global CIO of one of the top four U.S. pharmaceutical companies. He was interested in evaluating our robotics-enabled security service for deployment on their campuses in the U.S. and potentially abroad. This was not just another sales pitch; it was yet another make-or-break moment for our startup.

The demo was scheduled for Monday at 8.00 a.m. EST in Billerica, Boston. I arrived on Friday, roughly sixty hours ahead of the demo, to prepare on-site with one of my team members. Needless to say, there was a significant amount of work that went into weeks prior to this on-site meeting to ensure we were fully prepared on all fronts for this engagement. I was responsible for ensuring the demo went smoothly and that all our teams—hardware, software, and operations—were engaged on every front so that we could build trust and rapport with this gentleman and ensure our product addressed their core needs.

I arrived in Boston at around 7.00 p.m. and headed to the Billerica campus of this company. Once on campus, a teammate and I met our local point of contact for the facility. After a few handshakes with this gentleman, we began our drill.

The Challenge: Preparing for a Flawless Demo

For those of you who are into the startup ecosystem, you've probably noticed the difference between a demo or pre-sales meeting with an enterprise client versus a mid-sized client. A pre-sales meeting with a mid-sized client feels like a sprint or a 5K, while the ones with enterprise clients feel more like preparing for a marathon. There's a lot to factor in, and almost always, you need to address novel risks that emerge once you're on the customer's premises with your product.

Needless to say, we ran into multiple hurdles before the demo.

1. **Network Issues:** The facility's Wi-Fi was restricted using a captive portal and required our robot to go through a custom login process every couple of hours, so we had to write a script to automate this network re-authentication process.

2. **Cellular Coverage:** The location had no Verizon coverage, and our robots by default were equipped with Verizon SIM cards and the corresponding telecom module and firmware compatible with that frequency. This meant we had to do a last-minute switch to AT&T hardware and configuration on the robot. It also made it nearly impossible to communicate with our Bay Area team from Billerica, Boston.
3. **Operational Risks:** The site was filled with high-risk areas, including stairs, escalators, ramps, and a few places that looked like they were designed specifically to test a robot's will to survive hostile terrain. We had to map the entire facility and cordon off these zones on our robot's ROS map to prevent any unintended episodes in acrobatics during the demo.
4. **Non-stop Rehearsals:** Over Saturday and Sunday, we worked tirelessly—customizing the robot's hardware and navigation system, refining the demo script, rehearsing with remote specialists, and ensuring all edge cases were covered.

By Sunday night, we were brutally exhausted—but ready.

Our first in-person demo and meeting with the Global CIO were scheduled for the following day: Monday.

The Demo Day: Meeting with the CIO

At 6:00 AM on Monday, I arrived on campus. We conducted final rehearsals with robot specialists, backup engineers, on-site security teams, and facility personnel—essentially, with our remote team in the Bay Area.

At 8:00 a.m., the Global CIO arrived, shook hands with us, and was silent but highly observant for the most part. He observed quietly as we reiterated the problem statement and presented our solution, explaining how our robotics-enabled security service could address their pain points and elevate their security profile.

What he did not know? My teammate and I—two people troubleshooting one problem after another since we arrived on Friday—had practically lived at the facility for the past 60 hours to ensure nothing went wrong.

Post this first session, he asked two questions that defined the outcome of this meeting.

The First Question: 'Is Your Robot Secure?'

After the demo, the CIO walked up to me, looked me straight in the eye, and asked:

'Can you help me understand why this robot is fully secure? How can I trust it?'

I took a deep breath and laid out our security framework:

- **SOC II Compliance:** Our company is SOC II certified and has a Chief Security Officer responsible for audits and compliance.
- **Physical Security:** The Ethernet port is locked to prevent unauthorized access.
- **Data Security:** All data, both at rest and in transit, is encrypted using AES encryption and certificates.

- **Access Control (AAA Model):** Only authorized personnel can access the robot, since role-based permissions are in place.

He listened carefully. I could tell he was satisfied, at least for now. For those wondering how we secured our SOC II certification, we partnered with Vanta, a prominent security audit firm based in San Francisco. To say the least, they delivered deep value, and this engagement proved its worth. After answering these questions came the next arrow.

The Second Question: 'Where Do Your Cameras Come From?'

Then came the next question, sharp and direct: 'I see a lot of cameras on this robot. Where are they manufactured?'

I answered:

'Most of them are sourced from outside the U.S., given supply chain and pricing constraints.'

His follow-up was immediate: 'Do they come from (country) — a country he did not seem to trust at the moment?'

'Yes, sir. Today, most do.'

There was a weird silence in the room for a moment.

Then, he asked:

'Given the cyber risks, would your company be willing to replace all of them with cameras manufactured in the U.S.?'

I responded:

'Sir, I understand your concern and appreciate you bringing that up. With all due respect, as a startup, we must be mindful of our resources. Replacing the cameras involves major adjustments to our supply chain and pricing. If you are open to committing to a fleet of at least two dozen robots, we could make that change. I understand this may not be feasible right now, but that is what it would take for us to move forward. I shall have a discussion with my CTO and get back to you soon.'

He nodded respectfully, appreciating the honest and direct response.

When you are in the arena, you are always trying to balance the needs of the client with the mission and capabilities of your startup. It's challenging, but the truth is that, as a startup, time, talent, and capital are your most valuable resources. Every 'no' helps preserve them, and every 'yes' should be approached with caution.

Going back to the conversation, after he heard my response about the dozen robots and supply chain, he said something that immediately elevated the trajectory of this engagement:

'I see great value in your product. If you replace the cameras, I shall do my best to help you expand this service.'

We shook hands. He smiled—briefly. The meeting was over, but the opportunity had just begun. I carried this message back to the team.

Post-Demo: Deal On the Table

A few days later, we received feedback from our internal champion. The company saw real potential in our robots and expressed the interest to expand the fleet. They were convinced about our security posture but had two additional requirements: replace all cameras with U.S.-manufactured alternatives and ensure the robot could navigate elevators.

The camera replacement was a relatively easy fix; we sourced compliant alternatives.

But elevator integration?

On one hand, it was a significant technical challenge; on the other, we had a deal on the table. The prospect was willing to expand the fleet, and we had a deal if we could develop and present an MVP—not a full-fledged solution, but an MVP.

The Engineering Challenge: Elevator Navigation

Under our CTO's leadership, our engineers worked for weeks to integrate our robots wirelessly with the on-site elevators, specifically KONE elevators in this case. We achieved an 80% success rate; sometimes the robot hesitated and did not enter the elevator, but for the most part, it worked.

It is important to note that this integration was specific to KONE elevators, from what I remember. Elevator configurations vary sharply across manufacturers, and integrating with each would be a startup project in itself.

So, this was an important lesson — a reminder that working with Fortune 500 clients often comes with an array of customisation demands, strategic shifts, and trade-offs. You cannot afford to have all your clients in this bucket; too much custom work can slow you down. A healthy balance between Fortune 500 enterprises and mid-sized growing companies is key.

The balance depends on many factors, but the first is understanding the time and effort required for customization versus long-term ROI. In this case, we had buy-in for expansion, and an MVP—not a full-fledged solution—was enough. It worked, creating a win-win scenario for us and our partners.

So, if you're a founder negotiating with an enterprise client, it's best to secure a solid commitment before agreeing to customize your product. Even with a commitment, stick to your MVP and resist the urge to build a perfect solution. This approach alone will save you a ton of time and capital.

In this case, our engineering team built an MVP, we demonstrated it to the security team at this pharmaceutical company, and we sealed the deal. Our founders, from the get-go, kept a high bar on the talent, and that helped us unblock a wide range of technical hurdles that would have otherwise blocked sales.

So, to summarize:

In this case, we had SOC II certification, which helped us assure the CIO of our strong security posture. We secured a commitment to build an MVP for our prospect's elevator interface, accommodated their request to replace cameras with domestically developed ones and, most importantly, did so with a yet another commitment from them to expand the fleet of robots that would patrol this campus.

Overall, it was a win-win deal. As a startup, you have only so many engineers in the early stage, and they are all focused on ensuring your product is stable and is delivering sustained value to all your customers. Without capital, you can't afford to engage talent in projects like this unless there's a clear path to business expansion. Customization is almost always impossible to accommodate, so securing a firm commitment from your prospect before investing efforts is key.

Key Lessons for Startup Founders

This engagement taught us five important lessons about enterprise sales:

1. **Security Is a Deal-Breaker**
   If you are dealing with an enterprise client, security is not optional; it is the first filter. If you can't meet SOC II, GDPR, DORA, encryption, and access control (AAA) standards, it's going to be an uphill battle to even land a meeting. These companies have entire teams dedicated to rejecting vendors, and security is the most common reason—rightfully so. Take it seriously, or they won't take you seriously.
2. **Enterprise Clients Expect You to Adapt**
   Unlike startups or SMBs that might adapt to your product, enterprise clients assume you'll adapt to them. Some requests will push you in a direction you already wanted to go—those are worth it. If a lightweight MVP can meet a client's needs and benefit others, it's worth considering.

But if it demands a fully custom solution that turns you into an outsourced dev. team, walk away.

3. **You Do Not Need a Full-Blown Security Team**

   You don't need a full-blown security team to get your product enterprise-ready. We partnered with Vanta, a security audit firm based in San Francisco, and they delivered deep value. Being SOC II certified helped us extinguish many security fires that are prominent in these sales conversations, often within minutes. Needless to say, it added much-needed momentum to our enterprise deals.

4. **Your Hardware Tells a Story — Make Sure It's the Right One**

   Be mindful of hardware sourcing. Enterprise clients carefully evaluate every hop in your supply chain for long-term security and compliance risks. Plan accordingly, especially if you are in robotics. Your cameras, LiDARs, tablets, onboard computers, and pretty much all data-processing devices mounted on your robot will be scrutinised. They all tell a story to your client; make sure it is the right one.

5. **A Balanced Client Portfolio Is Critical**

   Big clients bring big cheques, but also long sales cycles, endless security reviews, and high expectations for support. If all your customers are like this, you'll spend all your time jumping through hoops instead of building your product. The best approach? Have a balanced mix of clients, so as an early-stage startup, you can keep your lights on and yet scale your impact.

The Next Win: Breaking into the Cloud Market

This experience reinforced our enterprise sales approach. Shortly after, we engaged with the number one cloud provider—an opportunity that converted into a major sales win, thanks to the combined efforts of our sales, engineering, and executive team. They all played a big role here, much like the Andes Mountains of Peru shape the course of the Amazon. More on this engagement in the next chapter.

Chapter 07: Meeting with a Cloud Giant at a Robotics Conference

This is one of those stories that reinforces how 99% of conferences and events that you will attend as a founder, or an engineer will feel like a grind and yield little to no impact on your sales. But then there's that 1% encounter—the kind that has the potential to completely change the trajectory of your startup. This is especially true if you are into robotics.

I saw this play out firsthand at the ISC West conference. In 2019, on the first day of this conference, we were in deep demo mode. Each of us was demonstrating our robotic-enabled safety and security service to people visiting our booth. It was mid-morning, around 11.30 a.m., and we had already spoken to and demoed our mobile robots to more than 100+ visitors. Some were curious, some amazed at what robots could do, and then there were a few we didn't expect—until they showed up.

At around 11.30 a.m., a small group of executives walked up to our booth.

Semi-casual outfits, flipped badges.

'What does your robot do?' one of them asked, standing a foot away from me and pointing at the robot.

Their badges were turned inside out, so I couldn't see their affiliations. I treated this conversation as we always do, with deep curiosity and a focus on the problem we were solving for our customers. As this conversation continued, more of their team members joined. The discussion deepened. From door and elevator integration to fleet operations and badge scanners, they covered it all. My teammates noticed and joined in as well.

After about a 20-minute-long, engaging exchange, they handed us their business card. That was when we realised this was not just any team. It was the number one cloud provider of our era — one of the two giants from Washington that shares its name with the world's largest river located in South America.

Put simply, this card was an invitation to partner with a Fortune 10 company — the global leader in e-commerce and cloud services. And just like that, a routine booth conversation turned into the beginning of what felt like an enterprise deal of a lifetime.

Over the next eight months, I witnessed some of the most profound stress tests conducted by a customer in the enterprise world. Here is how it all unfolded.

The Eight-Month Gauntlet

When a Fortune 100 company takes an interest in your startup, one of the two things can happen:

1. Endless custom integration requests.
2. A relentless stress test to see if your product can survive at scale.

We got the latter.

Over the next eight months, our robot went through a series of rigorous tests — designed not just to verify its capabilities but to identify its bottlenecks. These tests were designed meticulously by our customer to ensure our robot performed as expected and, more importantly, would cause absolutely no harm to people or their property by any means.

The following questions offer a glimpse into what these stress tests were designed for:

- Could our perception system detect something as fine as a thread?
- Could our robots operate reliably over a spectrum of lighting conditions, in noisy environments, and through long hallways?
- How well could our navigation stack handle unexpected obstacles?
- How reliable was the robot when handling staircases and elevators?
- What about escalators — would it roll down or detect and navigate around them?
- And what about battery life?

The list goes on...

During this time, every hardware and software component of our robot was meticulously examined and tested at a level that was beyond anything we had experienced before. It was a marathon full of challenges. These were some of the most gruelling tests our product had ever faced.

After all, how do you ensure your 3D Kinect sensor is able to detect a fine thread placed between two desks in office space? It is hard to detect it with the naked eye, let alone have a robot and its sensors detect it. This was just one of the hundreds of tests we had to pass.

The same applies to lighting conditions. Sun rays, internal office lights — you have to ensure your robot can navigate safely and detect obstacles under all lighting conditions. Truth be told, if a sensor like LiDAR can turn blind from a mere array of photons emitted by the sun, where do you even go from there? Sure, you can fuse data from different sensors and add navigation-restricted zones on the ROS map, and so on. Long story short, these tests were just a glimpse into what those eight months looked like.

On the first pass, we cleared a decent number of tests and had to rework our hardware and software to ensure we passed them all, while providing a timeline for the resolution of the remaining ones that were not a high priority in the short term but were in the longer term for this client. It felt like walking across a fire pit barefoot for all those eight months.

But some fires are not meant to burn you — they are meant to forge you and your team. They are meant to turn that coal into a diamond. That is what happened in our case. Through this period, which continued for about eight months, we made a list of all the hardware and software components of our robot that didn't stand this test and worked with our prospect to prioritize those that were non-negotiable for this deal.

What came next was essentially a contract—a deal to deploy a series of robots across their facilities in the U.S. Needless to say, we signed this contract. While it was not a large-scale rollout, it was a significant milestone—one that validated our work, the vision of our founders, and gave us credibility in this space. More importantly, it gave us the capital runway we desperately needed to fire our thrusters in the U.S. and beyond.

Over the next six months, we deployed about half a dozen robots across four facilities, helping this client identify safety and security risks using our robot-enabled security service. From unidentified guests visiting after hours to leaking sinks, unlocked doors, and unattended devices left on tables — our robots captured all of these security anomalies in real time. Given this performance, our client saw the value we brought to the table and decided to expand the fleet across several of their facilities throughout the U.S.

While this post might sound like a humblebrag about our startup, the real point I want to highlight here is this: if you're a robotics student or a founder, understand that there are both pros and cons to investing your time in trade shows and conferences. If you do decide to attend one, know what to look for — because there are always hidden gems behind those flipped badges. If you can spot them and spark the right conversation, you might just walk away with a potential offer of lifetime in your hand. That brings us to the lessons learned section for this chapter.

If you're a founder, team member, or startup advisor, here are key lessons from this enterprise deal that I hope will help you shape your startup's strategy around conferences and tradeshows.

Lessons Learned – The 1% of Conferences That Matter

1. **Treat Every Conference Like It Matters**

   Most conferences and events won't lead to game-changing deals—but some will. And the truth is, you never know which conversation will open that door. If you're attending SXSW, ISC West, or similar events, stay present, engage fully, and make every interaction count.

2. **Your Most Important Sales Pitch Happens When You Least Expect It**

   The best customers often won't introduce themselves as decision-makers. They observe, ask questions, and evaluate you without you even realizing it. Be ready. Always. In our case, their badges were flipped, and this team arrived in small groups—one half first, then the other—not all at once.

3. **Big Clients Bring Big Opportunities — But Also Big Tests**

   No matter how good your product is, make no mistake: Fortune 100 companies will not take your word for it. They will test, challenge, and verify every feature — rightfully so. If you can prove your technology meets their standards, it opens doors that no marketing campaign ever could. Be prepared for this grilling test. It is worth it.

4. **Credibility Is an Asset — Use It Wisely**

   Once you've worked with an industry leader, you don't need to advertise it directly — but that experience speaks for itself. It signals to other potential customers that your solution has been tested at the highest level. It will show in your team's spirit, actions, and, most importantly, their dedication to your mission.

5. **Enterprise Scrutiny Accelerates Product Maturity — Rapidly**

   Enterprise customers scale in ways mid-sized customers cannot, and that scale alone makes a difference. It's not easy—the demands are high, but the rewards are worth it. It propels you a hundred steps closer to your mission. And if the partnership fails, it exposes the weakest links in your system — whether in product, team, or operations. Either way, you learn and improve.

   In our case, it was both — a learning experience and a growth opportunity. We discovered that our Kinect sensors could not detect obstacles as thin as a thread, along with other edge cases related to staircases and escalators. At the same time, this partnership helped us expand our footprint beyond what we had imagined at the time. In the end, it was a win-win on all fronts.

Speaking of win-win deals, while securing Fortune 500 clients like this number one cloud provider through a conversation at a conference booth might seem like a good idea for startups, make no mistake: the long sales and product evaluation cycle might not be the right use of your time, especially if you are a startup running low on financial and human resources. So, what can you, as a founder or an engineer, do before you engage with Fortune 500 clients?

While there is no right or wrong approach here, the one we learned the most from boils down to your ability to identify clients that fall into the mid-sized company bucket and, among those, filter out the ones that are expanding organically at scale across the nation with a product that delivers real value. For us, we were able to strike a deal with one such prospect — a deal that ensured we did not *Slack*. Quite literally, as you will see in the next chapter.

Chapter 08: How a Strategic Partner 10x'd our Growth

Partnering with a fast-growing, mid-sized company can be a game-changer for a robotics startup, especially when that company is on a clear trajectory toward gaining industry dominance.

One such opportunity came when we partnered with a leading enterprise collaboration software provider that was expanding aggressively across multiple regions. Think about the biggest competitor to Microsoft Teams or Discord today, and then reflect on the explosive growth this mid-sized company experienced around 2019 before going public. This one was founded by Stewart Butterfield, and our paths crossed with the team here just as the company was about to go public.

Their need was simple but urgent: a safety and security service that could be seamlessly integrated with all their offices and scale alongside their business throughout the nation. The company was rapidly expanding its offices across the U.S. and abroad. It was preparing for its upcoming IPO. As a technology-focused company experiencing galactic growth across the country, they needed a security solution that would scale without creating operational overhead.

Our robotics-enabled safety and security service was a perfect fit in this case. Their founders and operations team were tech-savvy, and so were we. What started as a deal for just a few robots quickly scaled to more than a dozen within eight months—an unprecedented growth rate for a robotics startup operating in a security services domain. Back in 2019, almost no one was purchasing robots for security, so this deal was a major validation for us. That leads us to our first question: what was the exact problem we were trying to solve for this client? More importantly, why dedicate an entire chapter to this deal? We will cover each of these two questions one by one.

Cracking Product-Market Fit: Deploying Security Robots at Scale

At its core, the problem we were solving for this specific client was simple: a massive shortage of trained security professionals. If you are a tech-savvy startup experiencing galactic growth across the world, you tend to avoid getting into the hiring and firing process for security professionals and contractors. The reason is that you do not want to deal with the operational headache that comes with this process, and, to be honest, that is not your primary goal. So how do you stay focused on your mission of building the best chat app possible as a tech-savvy company and automate your safety and security services for your growing number of regional and global offices? The founders here decided to partner with us and address this problem head-on with robots. Our vision here was perfectly aligned.

Our startup addressed this gap by providing robotics-enabled safety and security services — robots that patrolled office spaces and warehouses autonomously using a hyper-tuned, ROS-based navigation system and flagged anomalies with a YOLO-powered image detection system. While I am not a roboticist, after having self-studied, built, and launched robot-enabled services for years across a few continents, I can assure you that our autonomous navigation stack was probably the most refined in the industry at the time. We had optimised it to handle thousands of edge cases, ensuring our robots navigated complex environments with near-flawless precision.

Given our alignment on vision and the sheer ability of our robot-led security service to address the pain point of our customer, this partnership had an inherent win-win scenario for both. That addresses the first question as to how this partnership started and what problem we solved for our customer.

Now the second question: why dedicate this chapter to this deal? There is a reason for that. This is not just an average mid-sized startup client. This is that mid-sized client that is experiencing galactic growth. They were opening half a dozen new offices every quarter across the globe. They were revenue-positive and heading towards an IPO. You cannot expect stars to align better than this if you are a startup and this is the client profile you have on hand. This deal alone is worth its weight in gold — if executed properly. These are super-rare deals. Hence, it deserves this chapter. So that naturally leads us to our next question: how did the growth trajectory of this startup fuel our own growth?

Well, in 2019, this client of ours was not only gearing up for its IPO but also rapidly expanding globally in cities like Denver, London, New York, Melbourne, and more. As they scaled, the demand for our robotics-enabled security services grew, and we were ready to deploy. That being said, no company would give you a green signal to deploy across all their offices in one go. They would test your product first at a few sites. The same happened here. One such deployment took place at their newly opened Denver facility — a large, high-traffic office space that needed a security robot to ensure seamless operations from day one.

Denver First — Then London and Australia

A colleague and I spearheaded the on-site deployment at the Denver office. After a strategic briefing from our sales team, I prepared the robot, conducted end-to-end testing, and validated all systems—software, sensors, and navigation module. We shipped the robot to the Denver office, and a week later, I flew in with another team member.

Once on-site, we met with our internal champion—a key representative from the client's team who reported directly to the decision-makers. Together, we conducted a thorough walkthrough of their security needs, identified critical patrol areas, and reviewed past safety incidents to shape our deployment strategy. Over the next 48 hours, we executed the deployment plan:

- **Precision Mapping & Navigation:** Developed high-accuracy ROS maps to power autonomous navigation system—ROS Planner.
- **Patrol Route Optimization:** Configured and programmed patrol routes to secure the perimeter while minimizing operational overhead.

- **System Calibration:** Conducted multiple test patrols, calibrating sensors and optimizing the navigation system.
- **Real-Time Threat Detection:** Demonstrated anomaly detection capabilities—hypertuned YOLO model.
- **Access Control Reinforcement:** Demonstrated our badge authentication service, ensuring only authorized employees could enter this facility after hours.
- **Real-Time Video Monitoring:** Tested the video call service with our remote operations team, enabling enterprise-wide oversight.

This deployment created the much-needed momentum for us as a startup. Our robot began patrolling the Denver office, generating weekly safety reports for the client and identifying safety and security anomalies that mattered the most to them. In short, it helped us demonstrate our value to them in a reasonably short amount of time.

Recognising that value, the client engaged us a few months later to expand operations into their offices in London and Melbourne, further embedding our solution into their global security framework. So, we did the same. In simple words, we were now on the global map just as this startup was, and this synchronicity effect that we became a part of was nothing short of magical because it gave us the momentum we needed across all fronts: revenue, credibility, and, most importantly, market validation.

The key takeaway here is simple: when you find a client that truly understands the value you're creating—and you hit that rare product-market fit—you've essentially discovered the magical runway of scale for your startup. This collaborative momentum continued to surge exponentially over the next six months as the company launched new offices in new cities, and we expanded right alongside them. It almost felt as if our growth curve was directly proportional to that of this high-growth, mid-sized company on its way to an IPO.

But while this might seem like the perfect runway for a startup, it comes with its own set of challenges. We experienced this first-hand.

Acquisition by Salesforce

Later that year, our client, after going through an IPO, was acquired by Salesforce. This meant that, if you were in our shoes, you would have to renegotiate all your deals with the new parent company. I have seen this happen multiple times — it is one of the biggest risks of working with mid-sized companies. If you negotiate well, an acquisition can open major doors for your product. But if you do not, you would have to start from scratch.

In many cases, the acquiring company already has established vendors providing similar services. Even if those solutions are outdated, inefficient, and come with operational headaches, big companies prefer working with fewer, larger vendors. That's where deals can go south. In our case, for this specific client, we lucked out. That is not always the case, though.

So, there are pros and cons to adopting this approach. But if you understand the runway, navigate it well, and position yourself strategically, you can build a business that thrives— even through acquisitions.

We have seen this story repeat with one more client, essentially a leader in the cloud infrastructure domain, who later was acquired by VMWare. In this case, luck was not on our side to say the least. While the outcomes and the trajectory of both acquisitions were different, the lesson remains the same.

If you have a list of prospects who fall into this bucket — mid-sized, high-growth companies marching toward an IPO and experiencing galactic growth, leading to your growth curve being almost directly proportional to theirs — in a way, you've hit a jackpot. But you also have to keep an eye on some of the death traps that, if not navigated carefully, can drain your startup. In summary, there are pros and cons to dealing with companies that fall into this bucket, and here are a few lessons that you want to keep in mind.

Lessons for Robotics Startups Working with Mid-Sized Companies

1. **Find Clients Who Deeply Understand the Value of Your Product**
   The best clients aren't just the ones with budgets—they're the ones who understand your value without you having to convince them. If they see the ROI early, they won't just buy your product; they'll build with you. The real advantage? They'll grow your business as they scale theirs.
2. **Speed is Survival**
   Mid-sized companies don't grow in a straight line; they expand in bursts. One day, they're testing your service in a single office. Six months later, they need deployments in five cities. The startups that win aren't just the best at what

they do—they're the ones ready to move when opportunity strikes.

3. **Acquisitions Are a Double-Edged Sword**
   A fast-scaling client is often an acquisition target. If they get bought, your contract might be on the chopping block. Position yourself as a critical service, not a nice-to-have. The best hedge? Make your value undeniable before the new leadership even arrives and evaluates all assets and contracts.
4. **One Customer is Not a Strategy**
   A dream client can be a trap. It's easy to pour everything into a single big customer account—until they change direction, slash budgets, or get acquired. The companies that last don't just build great products; they build resilience by diversifying their client base from day one. When you align with the right mid-sized, high-growth client, you become part of their expansion story. But that story can change overnight, especially if they get acquired. Be ready.

With this chapter, we've covered some of the most important lessons from my personal journey in robotics in Silicon Valley. But the book doesn't end here. We are now witnessing a tsunami of a new kind of robots—humanoids—that will soon become part of our daily lives. Over one billion humanoids are expected to inhabit Earth in the coming decade. This scale raises an important question: How should next-gen founders and engineers prepare for this era? What are the risks involved in building and launching humanoids at this scale, and what are some of the engineering trade-offs that come with them?

To answer these questions, we need to look closely at how companies around the world are building humanoids today. The lessons emerging from this space are equally critical for the next generation of robotics engineers and founders. We will cover these in the upcoming chapters.

Chapter 09: The Era of Humanoids

Automotive, chip manufacturing, and space exploration are the three areas where humanoids are being heavily adopted today. From BYD to Tesla, BMW to Audi, and Hyundai to Mercedes, automotive manufacturing plants are swarmed with humanoids.[12] Why humanoids?

If you've ever visited an automotive plant, you'll notice that the workforce requirements for these factories distill down to three core principles: mobility, efficiency, and repetition. You need workers who can navigate across factory floors often laden with parts and machines. You need them to carry these heavy parts and transfer them from one area of the factory to another. And, most importantly, this entire workflow, whether it involves automotive parts or transporting them from machines to conveyors or the inventory area, must remain both time- and resource-efficient. These are precisely the qualities that make humanoids such an ideal fit. You may ask: How?

If we take a step back and try to understand the anatomy of a humanoid, you'll notice that at a high-level, it is equipped with two legs that allow it to navigate diverse indoor terrain profiles: staircases, uneven surfaces, and a wide range of obstacles, resulting in an extremely high level of terrain mobility. Then it is equipped with two highly dexterous arms, or what we refer to in the robotics world as spatial manipulators. They enable spatial agility. This means humanoids can use these robotic arms to grab an apple or thread a needle through fabric with an extremely high level ofprecision. These two features, combined with the ability to teach humanoids to perform a wide range of repetitive tasks using modern AI models, add up to remarkable efficiency. Hence, humanoids are becoming an increasingly compelling solution in the manufacturing domain.

That naturally leads to this question: Which of the manufacturing giants are adopting humanoids, and how?

Global Adoption of Humanoids

From chip manufacturers to automakers and beyond, humanoids are being adopted across nearly every major industry vertical. In fact, it is no surprise that NVIDIA is partnering with Foxconn to deploy humanoids in its chip manufacturing plants.[13] After all, who doesn't want to automate the task of fabricating about 3.6 million enterprise-grade GPUs per year? Beyond NVIDIA, nine automakers around the world are already embracing humanoids on their production lines. This includes BMW, Mercedes-Benz, Tesla, Hyundai, Audi, Volkswagen, Nio, and BYD.

Need proof? Try visiting BMW's Spartanburg plant in South Carolina, here in the US, where you will witness Figure AI's humanoids working on the production lines without taking a day off for months in a row. Over a period of just 11 months in 2025, Figure AI's humanoids contributed to the production of more than 30,000 BMW X3 vehicles.[14] Beyond that, the team behind the Optimus bot has deployed humanoids on the factory floors of

Tesla. Optimus is currently being trained to automate assembly lines.

Similarly, Hyundai has adopted the Atlas humanoid developed by Boston Dynamics to automate its auto manufacturing plants.[15] In China, a local humanoid manufacturer, UBTECH, has secured partnerships to provide humanoid assistants to an array of automotive manufacturers including, but not limited to, BYD, Geely, FAW-Volkswagen, BAIC, and Foxconn.[16]

In Europe, Neura Robotics has launched its own humanoid to automate factories of Porsche and Mercedes.[17] Beyond automotive, the warehouse and logistics industry is also seeing a surge in the adoption of humanoids. Agility Robotics' Digit humanoid is being trained by Amazon to aid with logistics and last-mile deliveries. Trials are underway at Amazon's San Francisco training campus.[18]

Beyond manufacturing and warehouse teams, humanoids are equally being favored by household users as well. *Neo*, a home humanoid assistant being developed by the 1X team, is gaining popularity among those who dislike doing their laundry. For those concerned about affordability, Unitree has pushed the market forward by launching a humanoid priced under $5,000.

Given this adoption curve, some of you might ask: What about the production capacity? Isn't it extremely difficult to manufacture these systems at scale?

Production Capacity & Pricing

While the concerns around the production capacity of these humanoids are legitimate, we are now seeing this bottleneck being addressed at scale across the industry. To begin with, Tesla has confirmed a target production plan of about 100,000 for 2026. BYD, a Chinese manufacturer of humanoids that started with a plan for about 1,500 units in 2025, is now targeting 20,000 by 2026.[19-20]

While these targets may seem ambitious, make no mistake, they are well within achievable limits. Need proof? There's no better example of production scale than Agility Robotics, a company that has launched a dedicated production facility for humanoids targeting an output of 10,000 units in its first year.[21]

This surge in production is also rapidly driving down the pricing of these robots. For instance, the Tesla Optimus, currently priced around $30,000, is expected to see lower pricing as the company continues to expand its manufacturing capacity.[22] And for those still skeptical, look at China, where the production cost for a high-end, industrial-grade humanoid already hovers around $5,000 to $10,000.[23] That should give us a clear indication of where the market is headed.

Given this scale of production, the ongoing price drop, and the exponential adoption curve across industries, it's crucial to understand how humanoids actually work. To do that, we need to break down their functionality into the core systems that form their foundational layer. So what are these core systems? We'll cover that in the next chapter.

Chapter 10: Humanoid Foundations — Core Systems & Principles

If you take a look at humanoid robots today, you will notice their design and functionality can be distilled down to precisely seven core systems. If you understand these systems, you'll essentially have a solid foundation to build a humanoid that can solve real-world problems with efficiency, precision, and reliability. So, what are these seven core systems? We will now cover each of these, one by one, in this chapter.

7 Core Systems of a Humanoid Robot

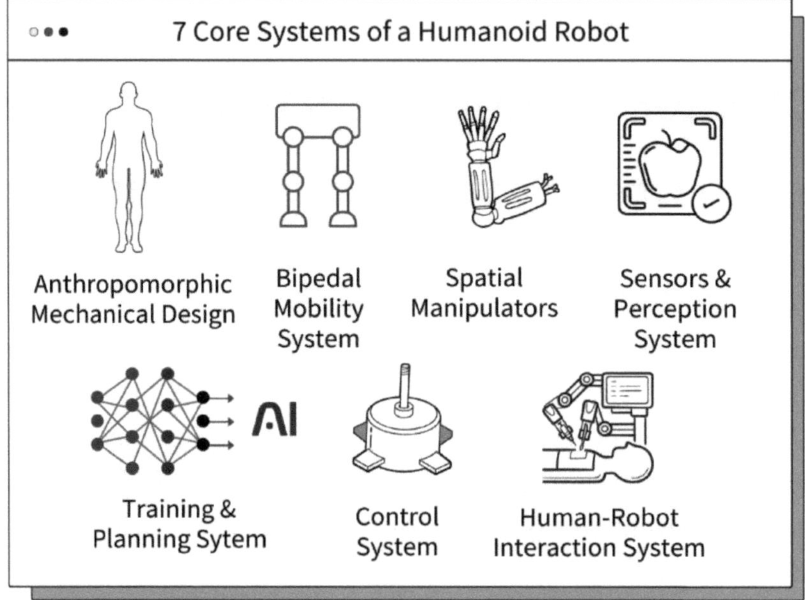

Figure 10.1 — 7 Core Systems of a Humanoid Robot

Illustration by the author.

Core System 1 – Anthropomorphic Mechanical Design

Well, the first one is anthropomorphic mechanical design, or, in simple words, human body-inspired mechanical design. At its core, a humanoid robot is built with a body plan inspired by our own: two legs, two arms, a torso, and a head. Each limb is made up of multiple joints, which are powered by actuators, i.e., high-power motors that allow the robot to walk, bend, jump, run, and handle objects. To keep it simple, we will distill this mechanical design into three sub-systems.

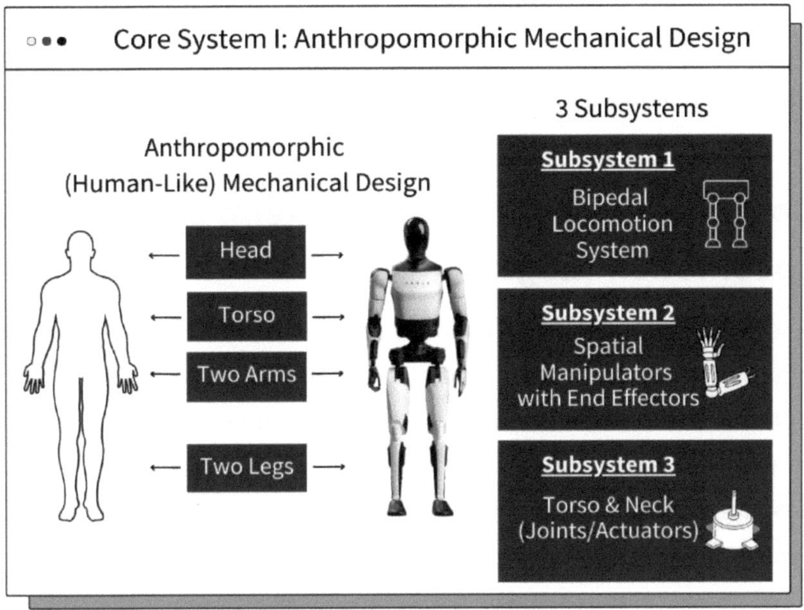

Figure 10.2 — Core System I: Anthropomorphic Mechanical Design

Illustration by the author. Tesla Bot image from Tesla "We, Robot" event (2024).

The first one is the bipedal locomotion system—essentially a fancy way to describe the two legs that each of these humanoids is equipped with. The second sub-system comprises the two spatial manipulators, or the two hands of the robot, equipped with end-effectors, or as some call them, grippers. This is the technical term for fingers with human-like dexterity. Then the third sub-system comprises the torso and neck. They play a key role in ensuring these humanoids have a wide range of motion while providing structural stability. This mechanical architecture is carefully engineered to balance strength, agility, endurance, precision, and reliability.

Now that we understand the first core system, which is the mechanical design, we need to understand the second core system, which is the application of the first core system. What do we mean by that?

Core System 2 – Mobility System: Bipedal Locomotion (Legs)

Just because a humanoid robot is equipped with two mechanical legs doesn't mean it is going to be able to walk and retain its balance. Locomotion is a right that a humanoid earns after it learns how to retain its balance. So, the question is: How does it do that? How does a humanoid robot ensure that at every step it takes, it doesn't fall?

Well, to ensure stability, a humanoid robot must understand the terrain profile it's navigating on, the slope of the terrain, the center of gravity of the robot itself, and then make a decision on what the next step should look like. This means that at each step, our robot has to take those parameters, namely terrain, slope, center of gravity, and more, as inputs and then make a calculated decision around how it's going to craft its next step. We call this gait planning in the world of robotics. Basically, it means determining what kind of footstep pattern the robot will choose for the next step based on the current parameters for the current step.

How does it do that?

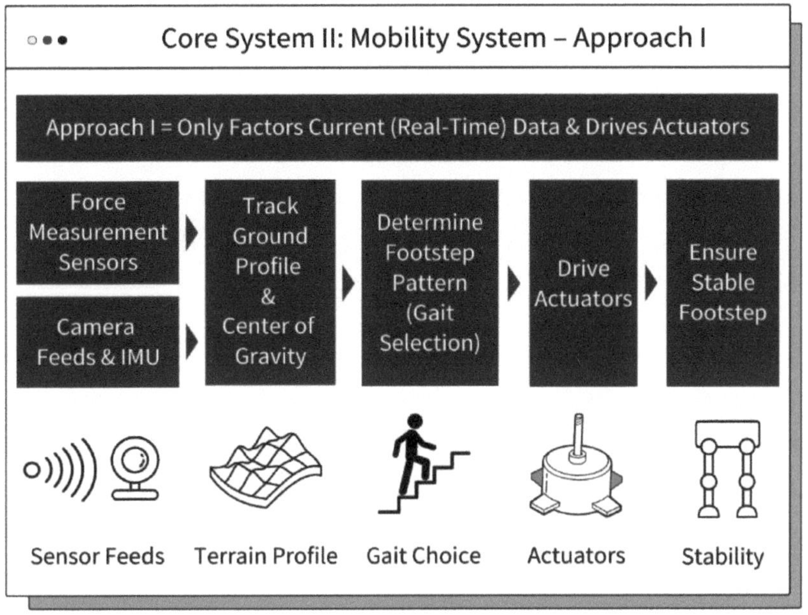

Figure 10.3 — Core System II: Mobility System – Approach I

Illustration by the author.

Well, there are two ways to approach this challenge. One is to simply take the real-time data generated from sensors located on this humanoid and measure both the ground profile and the centre of gravity of the robot itself, and then activate the corresponding actuators or motors located in the legs and the torso. This approach has been illustrated in Figure 10.3. These actuators apply appropriate torque, or, in simple words, the right amount of force and velocity, to ensure the robot maintains a firm grip on the ground. Thus, our robot is able to navigate on a given terrain from a source point to its destination point, while retaining its balance. Some refer to this approach as a reactive approach. Why? Well, this approach, if you notice carefully, only factors in real-time data received from sensors to make decisions. The challenge here is that if we only factor in the current state of the terrain and the robot and make decisions, we will be blindsided by any abnormal terrain profiles such as cliffs or slopes or pits that the humanoid may encounter on its way to the destination. Hence, this approach is reactive because we will end up reacting to those abnormal scenarios.

An approach that is more proactive, on the other hand, relies on the principle of prediction or, in mathematical terms, probability. So instead of just responding to the current step and the data measured from sensors at that moment, our humanoid scans the terrain, takes the current and historical data from the past few steps, and then predicts how the terrain is about to change. It then adjusts its next step or selects an appropriate gait, that is, type of footstep, based on these predictions. This approach has been illustrated in Figure 10.4.

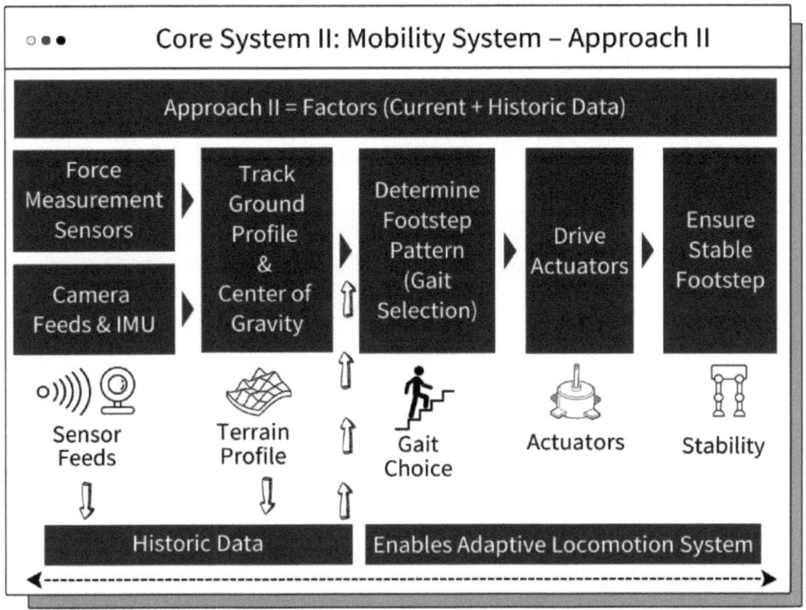

Figure 10.4 — Core System II: Mobility System – Approach II

Illustration by the author.

Long story short, this core system 2 is fully responsible for retaining balance and ensuring the humanoid can move through factories, warehouses, and homes with the highest levels of ground grip, precision, strength, and stability. Now you might ask: Well, that is for mobile navigation, what if our humanoid has to lift an object or pick a fruit from a basket? How does it do that?

This is where the spatial manipulators equipped with end-effectors come in handy.

Core System 3 – Spatial Manipulators & End Effectors (Hands)

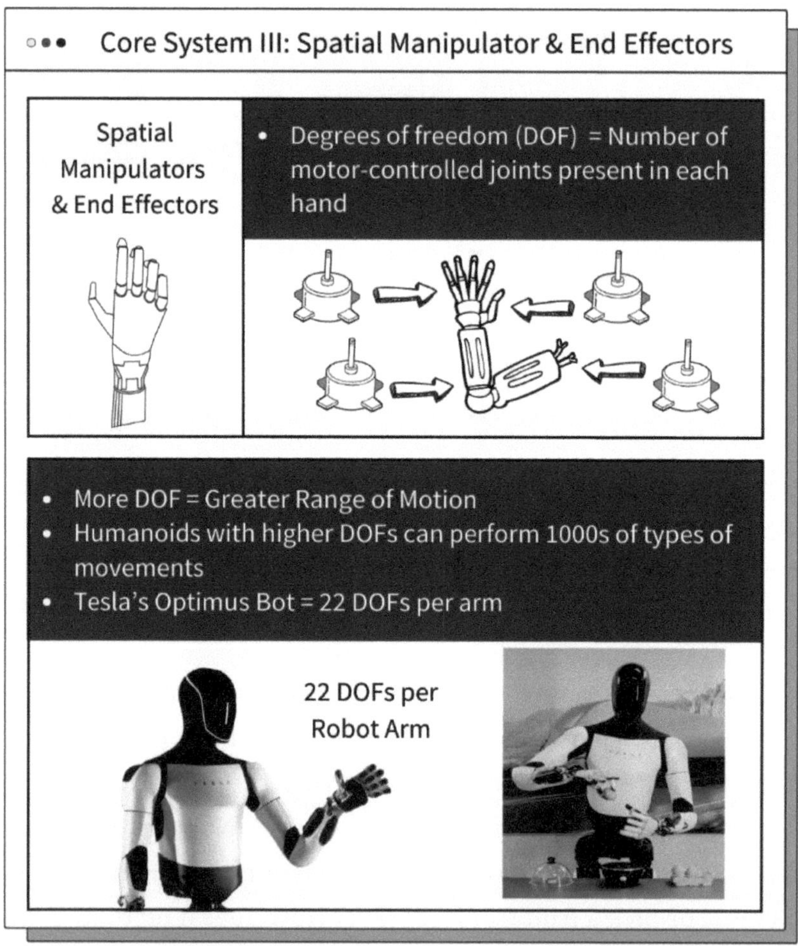

Figure 10.5 — Core System III: Spatial Manipulator & End Effectors

Illustration by the author. Tesla Bot image from "Optimus – Gen 2," Tesla YouTube channel (2023).

A spatial manipulator is a technical term used to describe arms with human-like dexterity, and an end-effector is a technical term used to describe fingers or grippers that can be controlled electronically to grasp an object.

Most humanoids today are equipped with two spatial manipulators, each equipped with strong end-effectors. From a mechanical perspective, these are essentially metal parts or limbs connected with high-torque motors that provide a range of motion. We refer to this as degrees of freedom in robotics, or DOF. So, if you hear folks say *3-DOF* or *4-DOF,* just know that means those are the number of motor-controlled joints present in each spatial manipulator. Why does that number matter?

Well, if a humanoid arm has more joints, that means each of those joints can be activated to move in different directions and at different speeds, and hence your humanoid can potentially perform thousands of types of movements. We humans have 7 degrees of freedom in each arm—that means we have 7 core joints. Three of them are located in the shoulder, one in the elbow, and three in the wrist. The same applies to humanoids.

While most humanoids today have 6–7 DOFs in each arm, Tesla's Optimus bot is one of the outliers with 22 DOFs in each arm.[24] That offers divine levels of dexterity with ultra-high precision and efficiency.

Now, we have covered the first three core systems, namely the mechanical design, the bipedal locomotion, and the spatial manipulators. The question is: How does our robot make decisions? How does it know how to move those spatial manipulators to pick an object? Objects come in different shapes and sizes. An apple is different from a car part that a humanoid may have to move on an assembly line.

Core System 4 – Sensors & Perception System (Eyes)

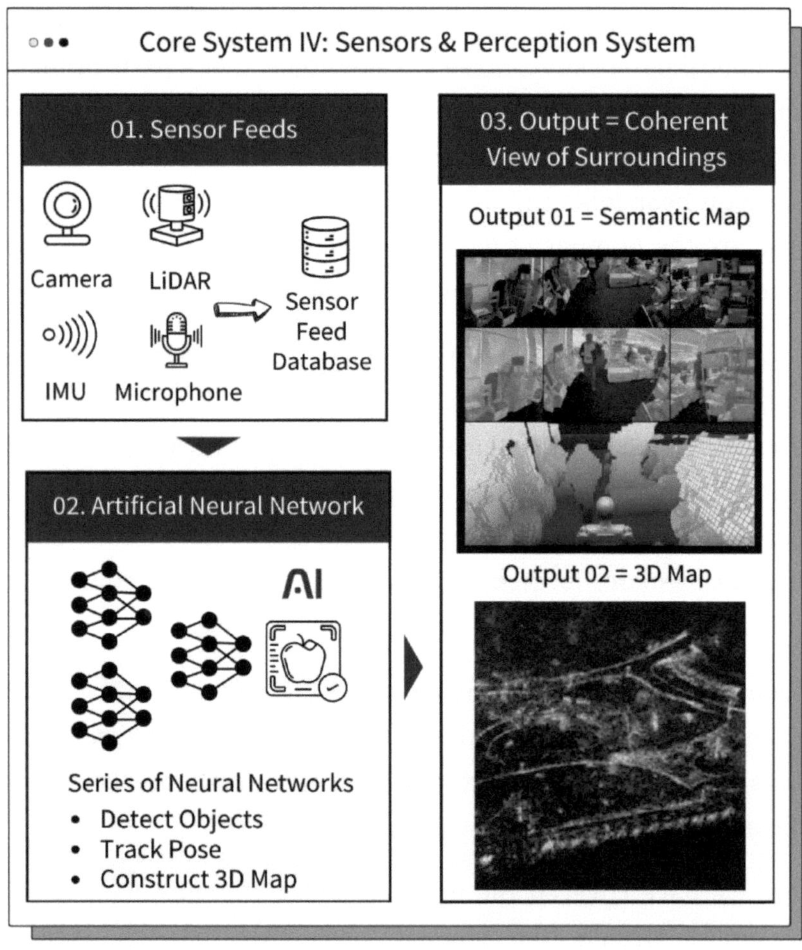

Figure 10.6 — Core System IV: Sensors & Perception System

Illustration by the author. Tesla Bot image from Tesla AI Day (2022).

So how does it make sense of these objects in the real world? That's where we need sensors and the perception system.

The fourth core system comprises sensors and the perception stack. What exactly is a perception stack? Well, we humans are blessed with biological sensors such as our two eyes, our ears, and our nose. Data captured by these sensors is processed by our brain, and that allows us to coherently perceive our surroundings. Similar is the function of the perception stack of the humanoid. A humanoid is typically equipped with multiple sensor modalities: cameras for visual recognition, depth sensors or LiDAR for 3D geometry, inertial measurement units for orientation, and microphones for audio input.

Data captured from these sensors is processed by the perception stack of our humanoid. This stack is simply a series of AI models or more specifically, artificial neural networks trained to detect objects, segment scenes, estimate 3D poses, and track motion of nearby objects over time. The output of this perception pipeline is a structured representation of the world, basically a real-time semantic map that tells the robot where things are and what they are.

This sensor and perception system is foundational to every modern humanoid because it is this vision-centric system that enables these humanoids to develop a coherent and high-precision understanding of their surroundings. Now, with this understanding of its environment, you might ask: How does it make decisions? How does it decide how to move its arms and legs to complete its task? That's where the next core system comes into play.

Core System 5 – Training & Planning System (Brain)

If you think about the array of tasks that our humanoid has to complete in this world, each of these tasks can be broken down into a series of actions, that is, sub-tasks that it can be trained on and optimized for independently. To better understand this principle, let's take an example.

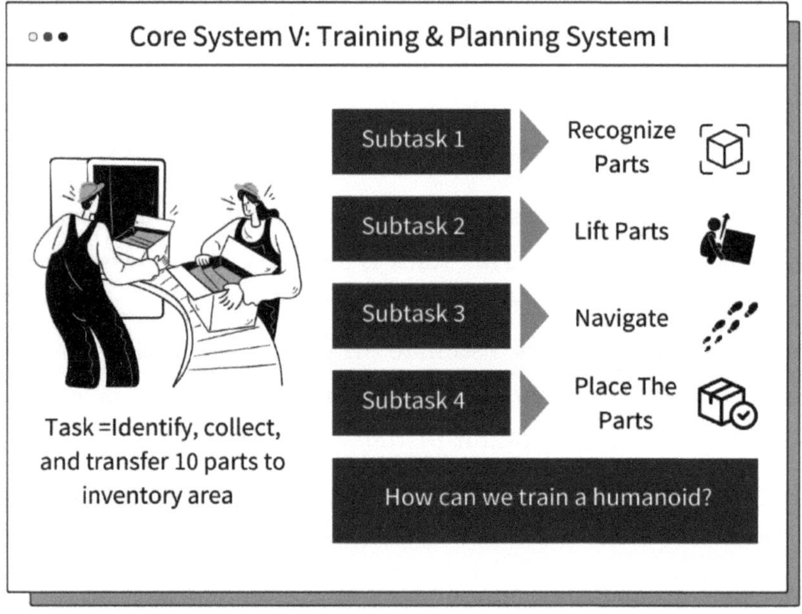

Figure 10.7 — Core System V: Training & Planning System I

Illustration by the author.

Let's say you have a humanoid deployed in an automobile factory. You want it to identify and collect ten parts that are released on the conveyor belt every ten minutes and transfer those to an inventory area in your factory. This task can be broken down into four sets of actions or sub-tasks as illustrated in Figure 10.7. The first sub-task in this case is to recognize the parts, the second is to lift ten of them, the third is to navigate to the inventory area, and the fourth is to place them there. This is obviously a simplistic way of looking at this task. The principle here is that each task is a combination of a series of actions or sub-tasks.

Now, if we need to train our robot to perform this task in a reliable and safe manner, we need to train it on each of these sub-tasks first and then on the task itself. So how do we do that? Historically, we used to hand-code robots with if-else loops. Those days are long gone. Modern humanoids rely on reinforcement learning to learn these tasks.

The way it is done is that, for each action or sub-task, there is a set of inputs and a required output. The input for each of these sub-tasks comprises two components. One is the data received from the sensors and the second is the goal itself as illustrated in Figure 10.8. Each sub-task has a clear goal.

For example, in the case of automotive factories, the first goal is to recognise the parts, that is, object detection; the second is to lift the parts; the third is to navigate the factory floor; and the fourth is to place the parts. Thus, for each of these sub-tasks, there is a goal that we want the humanoid to accomplish and that becomes one input, and the second input is the data from the sensors. So, these two inputs are provided to the artificial neural network. The required output is the action itself, such as picking an object.

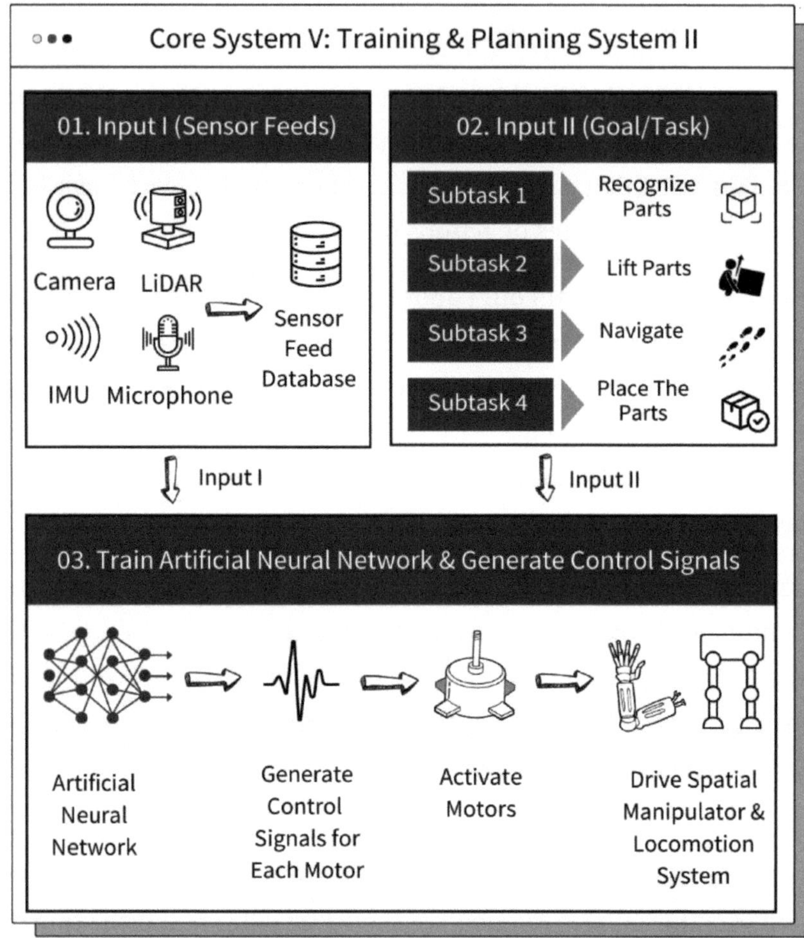

Figure 10.8 — Core System V: Training & Planning System II

Illustration by the author.

In simple words, the required output is the control signal or policies generated to drive the motors located around the body of the robot, both in its legs and its arms. So, the humanoid is trained using an artificial neural network over a large and diverse set of data until it consistently produces the required output.

For every accurate instance of output, this artificial neural function is rewarded, and for every inaccurate output, meaning if the object falls, it's not rewarded or, as some call it, negatively rewarded. Over a period of its training horizon, which could last for a few weeks or months, the humanoid arm gets really good at performing the given action. This process is then repeated for a series of actions, and finally for the end-to-end task.

The output of this process is essentially an artificial neural network that is fully trained to perform a specific task. Some call it the policy. Once trained on one robot, these neural networks can be deployed to thousands of humanoids at scale. Some of you might ask: Where do we train humanoids? In labs or in simulation environments?

This training can be conducted in the real world, i.e., a lab equipped with a physical humanoid and objects, or in a simulation environment. While there are pros and cons to each approach, it's this training process governed by artificial neural networks that offers humanoids the ability to perform tasks with speed, precision, and efficiency. This same training process can be used to teach a robot to walk without losing its balance, to climb stairs, to pick objects, and to perform all the other tasks you can think of in the real world.

This system of AI-enabled training allows humanoids to adapt to a series of scenarios that it might have to face in the real world. Now you might ask: Well, now that it has been trained to perform that task, how does it actually translate the output of these AI-models, i.e., intended action, into motion? How does it move its arm and legs in alignment with the specific action?

That's where the next core system comes into play.

Core System 6 – Control System (Action)

The sixth core system is the control system, which you can think of as the robot's nervous system. This is the layer that takes all the high-level plans and policies dictated by the artificial neural networks used to train humanoids and translates them into precise motor commands that drive motors located in the joints of the humanoid. Thus, they turn those decisions dictated by the artificial neural networks into actions.

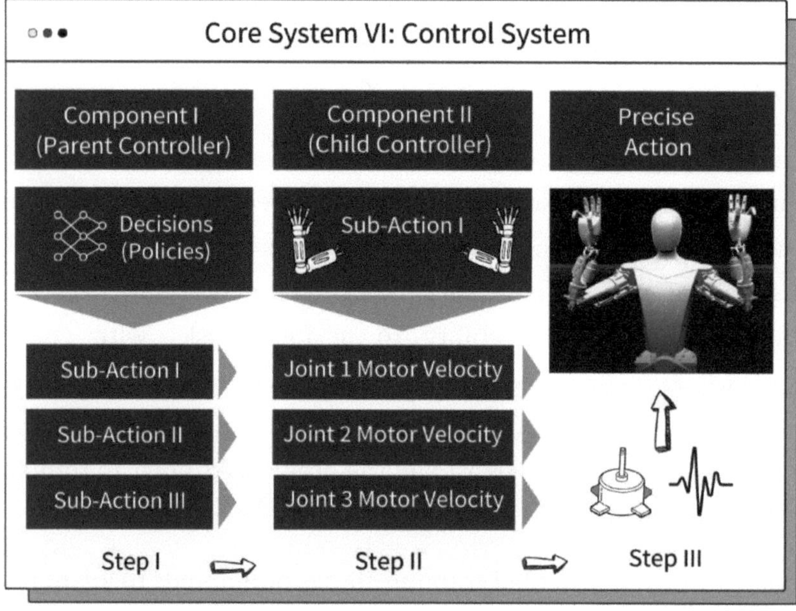

Figure 10.9 — Core System VI: Control System

Illustration by the author. Tesla Bot image from Tesla AI Day (2022).

This system is composed of two core components, as illustrated in Figure 10.9. The first one is the whole-body controller, sometimes called the parent controller. This controller is responsible for taking the decisions from the neural networks and breaking them down into sub-decisions or sub-actions dedicated to each of the hundreds of individual controllers that drive the joints located throughout the humanoid.

The second component is the collection of individual controllers or motors that are located in each joint present in the arms and legs. We refer to these as the joint controllers. They are responsible for managing the torque, velocity, and position of each joint for a given task. Together, the parent controller and the child controllers, or joint controllers, translate the intent finalized by the artificial neural networks into precise actions for the humanoid.

Now that we understand the top six core systems that are foundational to each humanoid, we still need to ensure that all of these actions that our humanoid will perform need to be executed in harmony with nearby humans. What do we mean by that? We humans operate with a certain set of principles and values, and we want to ensure that since humanoids will be operating around us, they have to be trained on those sets of values and principles. This is where the next core systems come into play.

Core System 7: Human-Robot Interaction System

While, in theory, a humanoid can be trained to move parts in a factory at galactic speed, that might yield ultra-high production capacity and throughput, but it comes at the expense of human values. Any action taken to the extreme yields extreme results, and this is both good and bad.

On the positive side, that means higher production in factories. But on the negative side, that also means ultra-high risks. What if a humanoid goes rogue? What if it continues to perform at that speed even when the other systems around it such as the conveyor belt stops working? What if it keeps operating at that speed and doesn't realize a nearby human worker is exhausted and about to fall on the ground?

You see? We humans, when we work, whether at our office, in a store, or in factories, we are not just performing an action. We are cognitively aware of our surroundings, and that is essentially our consciousness. We need humanoids that are not just great at performing tasks but also adhere to a certain set of values and carry with them a level of awareness. Those two aspects are probably the hardest to execute in this industry. Why? Those are metaphysical aspects that make us human, and replicating them on a machine has, at least to date, been an extremely challenging task, to say the least. We need to ensure that humanoids are aware of and trained on the values and principles that are core to our biology and evolutionary cycle. That is precisely the function of the human–robot interaction system, illustrated in Figure 10.10.

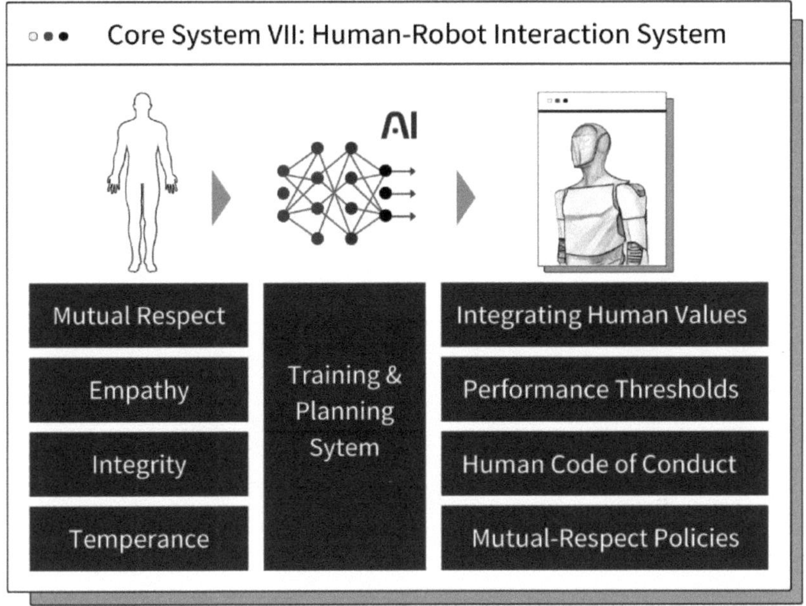

Figure 10.10 — Core System VII: Human-Robot Interaction System

Illustration by the author.

While there are a series of experiments and research efforts being conducted to explore the possibility of adding consciousness into these humanoids, we are still at the early stages here. For now, one thing is clear: we are approaching the much-needed levels of dexterity required to perform actions in the real world with humanoids, but we are probably decades away from understanding human consciousness with surgical depth and figuring out ways to embed that level of awareness with precision and depth into humanoid robots.

Speaking of human-robot interaction systems, now it's time for us to understand how these humanoids and other modern robots are being developed by leading players in the industry. More specifically, it's important that we learn the approach adopted by these world-class teams and the lessons we can learn from them as we continue to build the next generation of robots.

Chapter 11: Decoding Tesla's Approach to Building Humanoids

If you trace back the roots of Optimus, the team at Tesla first revealed the concept version around 2021. The core vision behind this project centered around the concept of a world built by humans for humans. What does that mean?

Optimus Bot - Vision & Evolution

Well, the underlying debate here is that, of all the tasks that we do today as humans, there are a significant number of them that are boring, repetitive, and provide minimal nourishment if any, to our mind. Some of these are even dangerous. The team at Tesla wanted to address this challenge by building a general-purpose humanoid that can perform these tasks and free up time for humans to focus on high-value tasks. So, around 2021, they launched the concept version that you see in Figure 11.1.

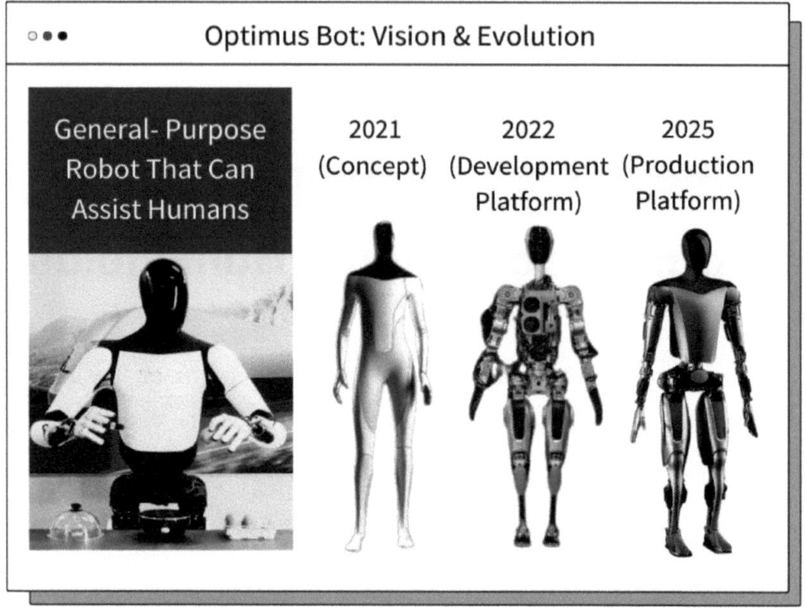

Figure 11.1 — Optimus Bot: Vision & Evolution

Illustration by the author. Tesla Bot images from Tesla AI Day (2022) and "Optimus – Gen 2," Tesla YouTube channel (2023).

If you carefully observe the evolution of this humanoid's overall form factor illustrated in Figure 11.1 and the corresponding specifications listed in Figure 11.2, you will notice that a reasonable number of parameters are aligned with those of an average human body. Speed, its capacity to lift weight, and so on are some examples of this phenomenon. However, there are specific areas where there is no doubt this humanoid greatly outpaces human capabilities, thanks to the underlying engineering and the design developed by the team at Tesla.

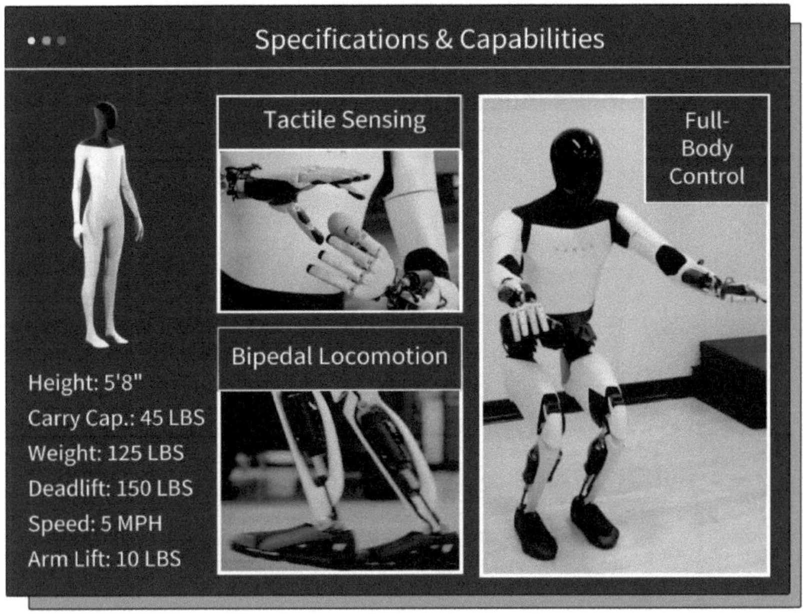

Figure 11.2 — Specifications & Capabilities

Illustration by the author. Tesla Bot images from "Optimus – Gen 2," Tesla YouTube channel (2023).

Over the period of the last five years, this team has made tremendous improvements both in the mechanical design, actuators, spatial manipulators, battery, hardware, and the underlying software used to train these robots. Their approach has significantly elevated the capabilities of Optimus. Today, you can see many of the latest versions of Optimus deployed in Tesla factories. Given this pace of innovation and scaled-adoption, it is prudent that we understand the underlying core principles that govern the engineering of this humanoid robot.

Optimus Bot - 6 Core Systems

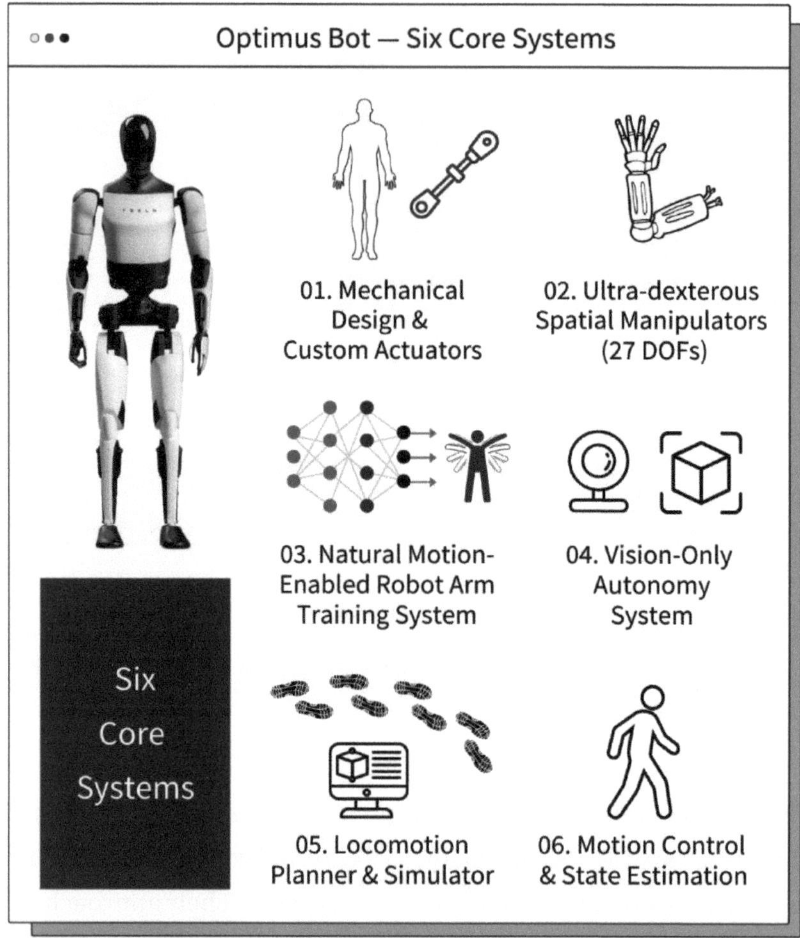

Figure 11.3 — Optimus Bot — Six Core Systems

Illustration by the author. Tesla Bot image from the Tesla "We, Robot" event (2024).

If you observe the Optimus Bot carefully, you will notice that, from an engineering and design perspective, it is composed of six core systems, as illustrated in Figure 11.3. This includes, but is not limited to, the mechanical structure and custom actuators, ultra-dexterous spatial manipulators with 27 degrees of freedom, bipedal locomotion and planning system, compute engine and the neural networks, sensors and navigation system, locomotion and planning system, and, last but not the least motion control system.[25]

Now, we will cover each of these systems one by one.

Core System 1 - Mechanical Structure & Custom Actuators

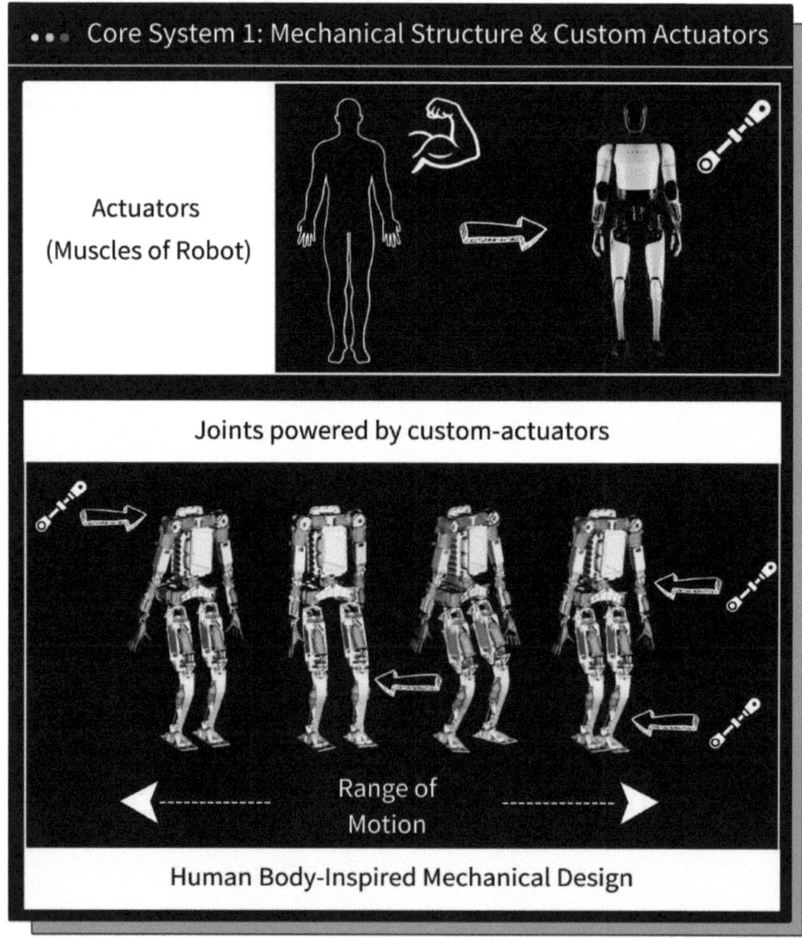

Figure 11.4 — Core System 1: Mechanical Structure & Custom Actuators

Illustration by the author. Tesla Bot image from Tesla AI Day (2022).

When you think about building a humanoid robot, you want to start with the required muscle strength of the robot. Why? Well, a humanoid robot by nature is intended to do a lot of physical work. This means its ability to perform that work with precision and efficiency depends on the quality of its muscles, i.e., actuators, as illustrated in Figure 11.4. The actuators are the muscles of the humanoid robot, and you can't get very far without getting them right. You might ask: what does an actuator look like in the real world? Think about it like a high-torque motor. So, what do we seek in these actuators when we build a humanoid?

We need actuators that are powerful, compact in size, support high-precision operators, and are power-efficient, all at the same time. If you ask a mechanical engineer to design an actuator with those constraints, you'll realize how hard of a problem that is. The team at Tesla understood this mechanical bottleneck from day 1 and decided to address it head-on. Through the initial experiments conducted by this humanoid team, the team came to the conclusion that off-the-shelf actuators available in the stores won't be a good fit for the humanoids as they were not capable of meeting those requirements. So, what did they do? In a classic Tesla way, they designed a portfolio of custom actuators in-house that could meet those requirements.[25] How did they do that?

Using Simulations to Determine the Right Actuators for Building Humanoids

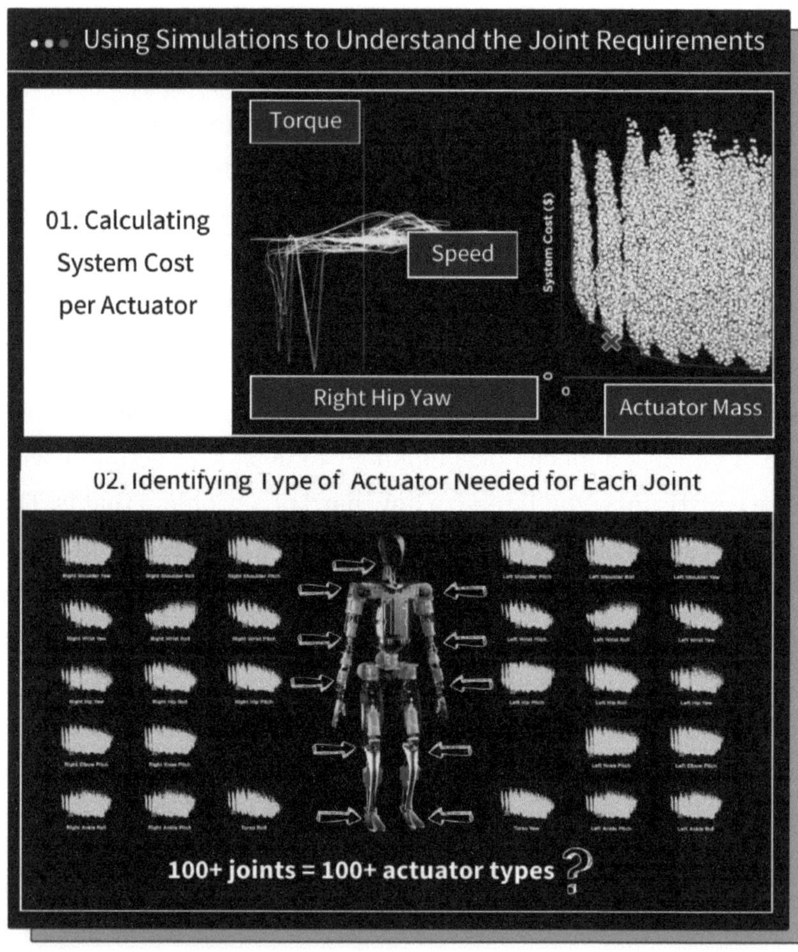

Figure 11.5 — Using Simulations to Understand the Joint Requirements

Illustration by the author. Tesla Bot image from Tesla AI Day (2022).

They first simulated the different types of movements a general-purpose humanoid like Optimus will have to execute in factories and warehouses. Then, using these simulations, they extrapolated the speed, torque, and power requirements for each of the joints of the humanoid. The challenge they ran into at this point was that, if a humanoid has 100+ joints, it will need 100+ types of custom-designed actuators using this approach. That's not a scalable solution, especially if you are planning to manufacture humanoids at scale. It's a production nightmare.

So how did they address this challenge?

The team approached this challenge with a classic optimization-styled solution, as illustrated in Figure 11.6. They simulated every single joint this humanoid needs, they mapped torque requirements for each, clustered them into groups using classic iterated optimization processes, and reduced the type of actuators needed to build Optimus from 100+ down to mere six: three rotary actuators and three linear. Only six custom parts to power 100+ joints.[25]

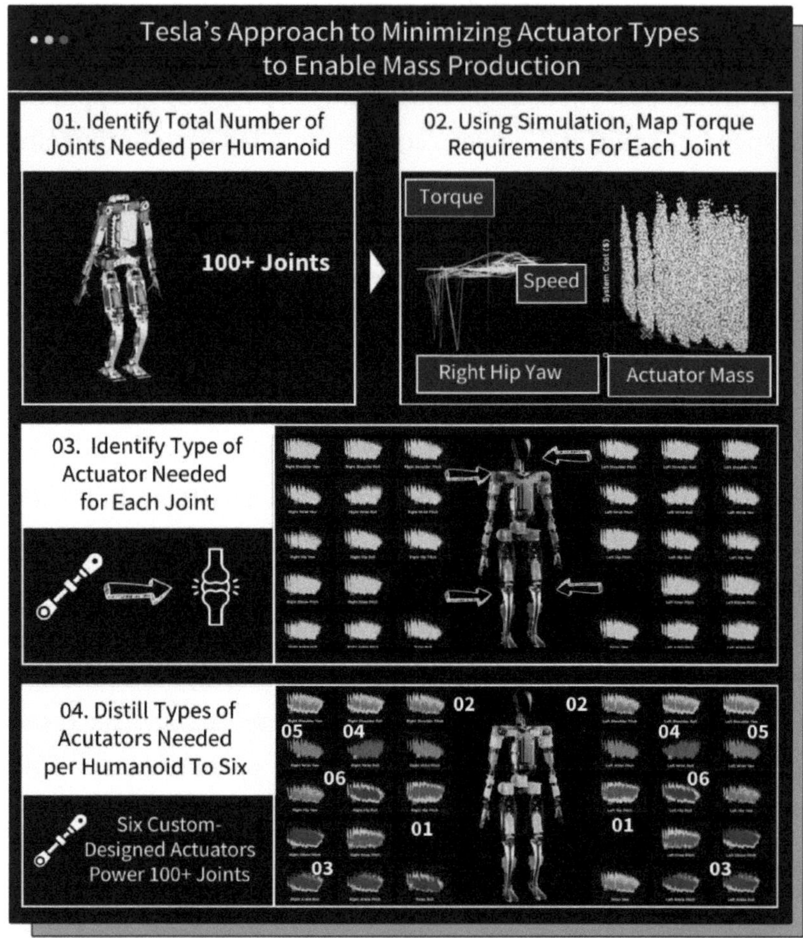

Figure 11.6 — Tesla's Approach to Minimizing Actuator Types to Enable Mass Production

Illustration by the author. Tesla Bot image from Tesla AI Day (2022).

Fewer parts, faster builds, galactic production speed. This project is indeed a masterclass in engineering. So, how powerful are these actuators in real life?

These custom-designed actuators by Tesla are extremely powerful and versatile, and that's why they were able to demonstrate a single actuator lifting a 500-kilogram grand piano during a live-streamed demo a few years ago. That combination of strength and precision is a big plus for anyone building humanoid robots. Why?

Humanoid robots use at least two dozen of these actuators, and they are responsible for enabling strength, precision, and stability in these machines. They are the muscles of the robot. Meaning, if they are not designed properly, you would have a hard time ensuring the robot stays stable and upright as it performs its operations, let alone ensuring those operations are precise and efficient. Some of you might ask: Well, doesn't each joint in the humanoid robot, whether it's the knee, the ankle, or the wrist demand different torque requirements? That's correct. So, don't we need different types of actuators? That's correct.

6 Custom-Designed Actuators

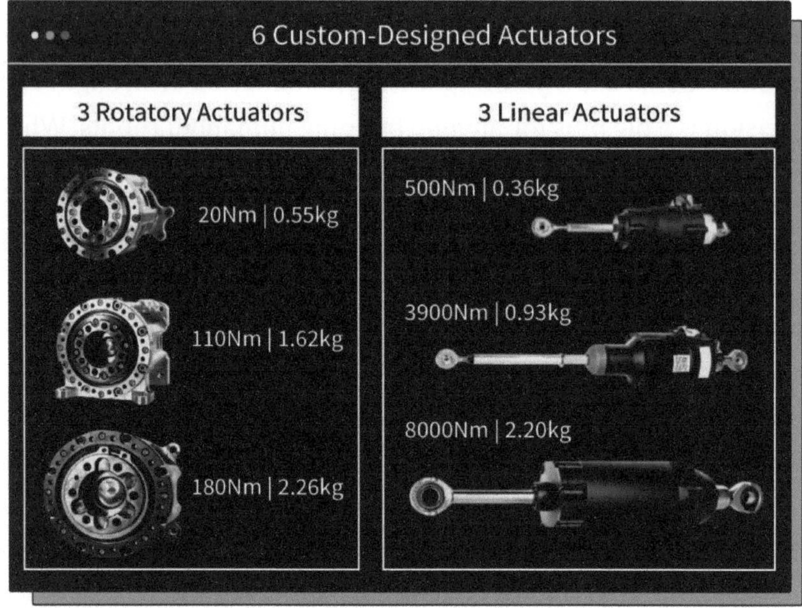

Figure 11.7 — 6 Custom-Designed Actuators

Illustration by the author. Actuator image from Tesla AI Day (2022).

That's the very reason Tesla designed six classes of actuators, each catering to the specific needs of strength, agility, and precision, or, to sum them up using one word, *torque,* for a wide range of joints present in the humanoid, as illustrated in Figure 11.7. The low-torque ones (20Nm) are used in the wrists and elbows where we need extremely high-precision and low-torque. The medium and high-torque ones, up to 180Nm, go into the hips and ankles because those joints carry most of the load. To carry heavy weights or parts, the robot's knees and hips also have linear actuators that can provide an additional lift of up to 8,000 N.[25] That's roughly equivalent to supporting several times the robot's own weight while ensuring the robot remains stable and upright even on uneven terrain.

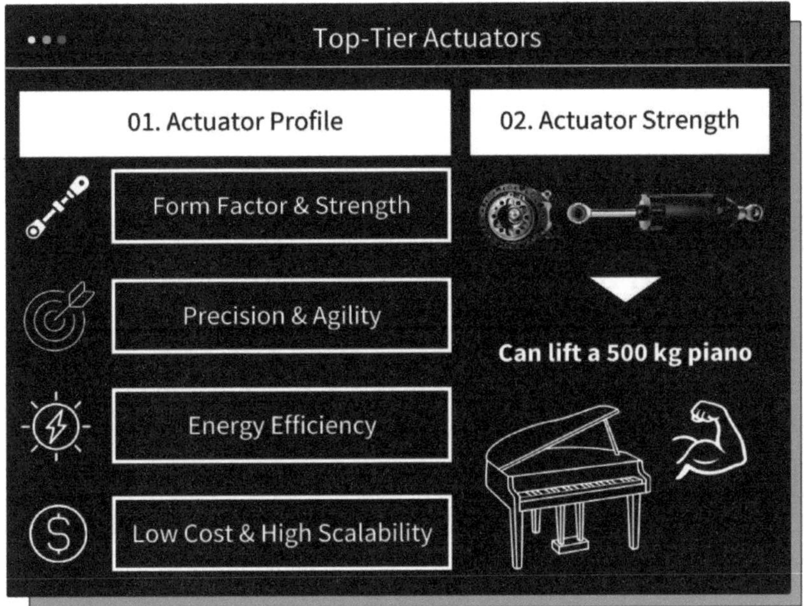

Figure 11.8 — Top-Tier Actuators

Illustration by the author. Actuator image from Tesla AI Day (2022).

But raw strength isn't enough. The robot needs to know exactly how each joint is moving at every moment. That's why Tesla integrated high-resolution encoders that track the precise angle of each actuator and torque sensors that measure the force exerted by each of those actuators in real-time. This means if Optimus encounters a potential fall or a resistance, it can proactively recover from such failure modes because it's able to analyze the data generated by the corresponding actuators involved in executing that specific task in real-time and adjust the corresponding inputs supplied to them. Thus, the approach adopted by Tesla to custom-design novel and top-tier actuators has resulted in extremely high stability, precision, and energy efficiency for these humanoids.

Some of you might ask: Beyond the actuators, what about the body of the robot? The frame? How did this team ensure the body itself is strong enough?

Robot Frame Design: Aerospace-Grade Aluminium and Lightweight Polymers

The team at Tesla used two core elements to build the robot's frame: aerospace-grade aluminium and lightweight polymer composite panels.[25] Why do we need these two elements? Well, if you think about it, there are parts in the humanoid robot that have to bear the most amount of stress, and those parts are made of aerospace-grade aluminum. The robot arm, for example, has to carry a lot of weight and hence is made of this element.

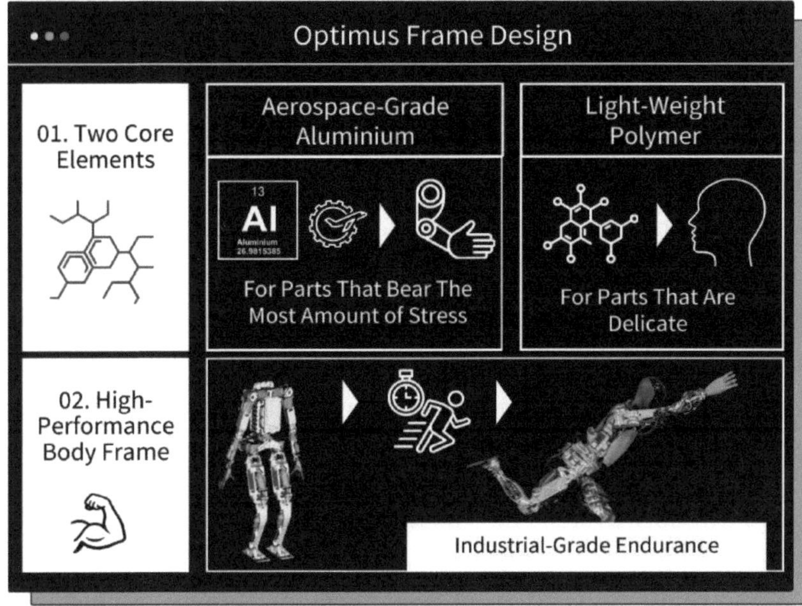

Figure 11.9 — Optimus Frame Design

Illustration by the author. Tesla Bot image from Tesla AI Day (2022).

On the other hand, there are parts that don't quite have to bear as much stress but they do need to be reasonably strong, and that's where the lightweight polymer composite panels are used, as illustrated in Figure 11.9. The head of this humanoid is a good example of parts that can be designed using lightweight polymers.

Thus, this combination of aluminium and polymer composite results in a design that is strong, lightweight, and extremely power-efficient. All of this engineering adds up to a staggering 200+ degrees of freedom (DOFs) in the body and around 27 DOFs in each hand alone. You may ask: What are DOFs?

A DOF, in simple words, corresponds to the number of independent joints present in the robot. Why is this number significant in robotics? Well, it's this number that governs how many different ways a robot can move a specific joint or a part. For example, we humans have only 7 degrees of freedom or 7 core joints in each of our hands. The latest version of the Optimus Bot, on the other hand, has about 27 DOFs in each arm. That means the Optimus Bot has an arm that is at least four times more versatile than a human hand, and hence it can perform a wide range of operations with precision and efficiency that a human hand may not be able to perform.

So, it's this core system, i.e., the mechanical design, that allows the Optimus robot to perform its operations and navigate in the factories with strength, precision, efficiency, and dexterity. It forms the foundation on which the team deploys the remaining six core systems. Now that we have seen the first core system, the next major one is the ultra-dexterous spatial manipulators, or, in simple words, the arms of the robot.

Core System 2: Ultra-Dexterous Spatial Manipulators

One of the biggest challenges in building a humanoid robot is designing robotic hands, or spatial manipulators, that can actually perform tasks in the real world with human-like precision and efficiency. Why? Most industrial robots that are operational today use conventional robotic arms, i.e., spatial manipulators that rely on grippers or suction cups to grasp objects. The problem with these grippers and suction cups is that they can only handle a limited range of tasks, such as picking an object that has an extremely small form factor, such as a strawberry, or rotating a door knob. A vast majority of these are designed for dedicated tasks and object profiles. The challenge here is that, if you really want to build a general-purpose humanoid robot, you need far more agility and precision than these arms will ever provide. That is a big problem. You may ask: Why?

Optimus is designed to work in human environments, which means it has to be able to handle objects of all sizes, shapes, and fragility. It should be able to handle items that are as fragile as an egg or glass to automotive parts that come in all shapes and sizes. Given these requirements, it's obvious that the existing set of robotic arms will not do us any good. Beyond agility and precision, there is one more challenge associated with the existing designs of robotic arms. Any guesses?

The existing approach of building a robot arm is simply not scalable. If you think about it, conventional robotic arms require one actuator per joint, which means one motor per joint.

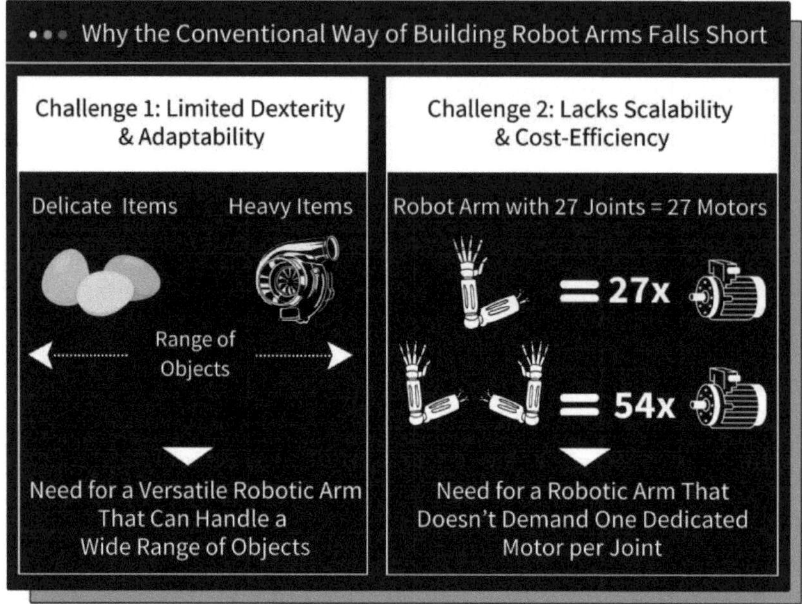

Figure 11.10 — Why the Conventional Way of Building Robot Arms Falls Short

Illustration by the author.

Now imagine a robot with 27 joints per arm. That's 54 joints in total for two arms, meaning you would need 54 motors to build them, as illustrated in Figure 11.10. Tesla plans to produce around 100,000 of these humanoids, so clearly, this architecture is neither scalable nor cost-efficient, and it doesn't address the use case under consideration.

So how did Tesla solve this problem?

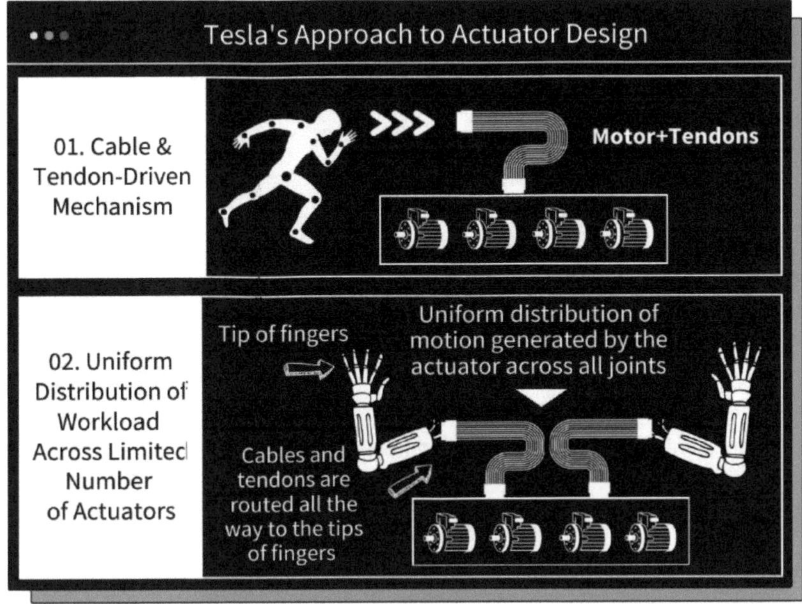

Figure 11.11 — Tesla's Approach to Actuator Design

Illustration by the author.

In their classic innovation-first style, they developed a novel design for a robotic arm from scratch to ensure it met those requirements, as illustrated in Figure 11.11. Tesla developed what's called an Underactuated Hand Architecture.[25] *Underactuated* means there are fewer motors (actuators) than joints, so instead of assigning one motor to drive each joint independently, the hand uses a smaller set of actuators connected by cables and tendons to control a series of joints present in parallel. Some of you might ask: How many actuators is Tesla using per robotic arm?

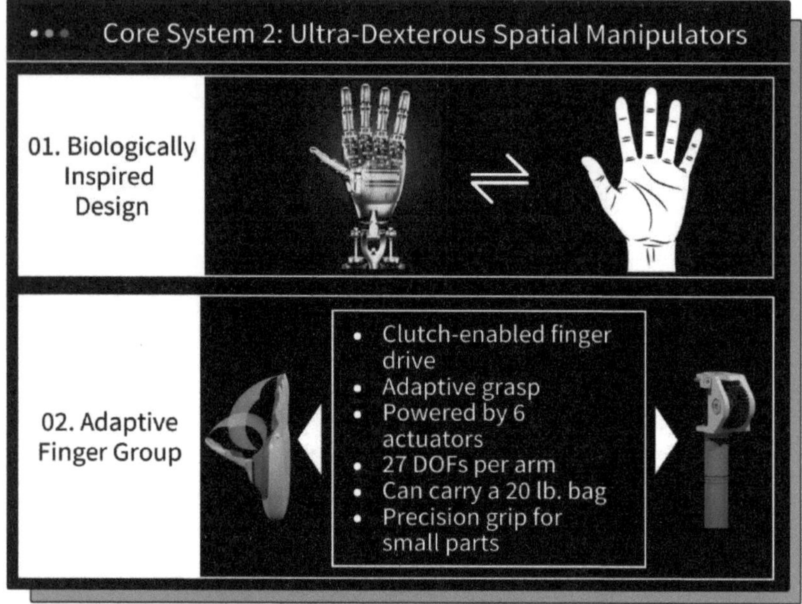

Figure 11.12 — Core System 2: Ultra-Dexterous Spatial Manipulators

Illustration by the author. Tesla spatial manipulator image from Tesla AI Day (2022).

In the case of Optimus, each hand is equipped with about six actuators driving 27 degrees of freedom, or about 27 joints. This approach of distributing the workload of multiple joints across the same set of actuators is only made possible because of a cable-driven mechanism. Inside each hand of the Optimus, tendons and cables are routed all the way up until the tip of the fingers and they are connected to those core set of actuators. This approach ensures that the motion generated by each actuator is mechanically and equally distributed across all the joints. This approach enables operational and mechanical efficiency.

Here's how it works in practice: When the robot wants to grasp an object, the six motors located in that hand start pulling on the tendons inside the hand. These tendons run through the fingers and connect to multiple joints along the way, as illustrated in Figure 11.11. If one finger touches the object before the others, it stops bending due to the resistance generated by the surface of the object in contact with the finger. Meanwhile, the other fingers, which haven't made contact yet, keep moving until they also touch the object. The tension in the cables automatically spreads itself out, letting the hand naturally adapt to the shape and size of the object it's about to grasp. This process is called passive distribution of tension, or passive compliance. It means the hand can conform to a wide range of object profiles without requiring a dedicated motor in every joint or complex software to calculate the perfect finger positions.

This design is both operationally efficient and robust. Beyond operations, this approach also requires fewer parts per robotic arm. This makes the arms much lighter and easier to manufacture at scale compared to their conventional counterparts that need dozens of motors.

Once trained using an AI model, this 27 DOF robot arm achieves ultra-high levels of dexterity. From needles to auto parts, it can grasp objects of all shapes, sizes and forms.

Now that we have covered the first two core systems, some of you might be wondering: How did the team at Tesla design this ultra-dexterous robot arm? What was the approach they adopted? This is, in fact, the job of the third core system, i.e., Natural Motion-Enabled Robot Arm Training System.

Core System 3 - Natural Motion-Enabled Robot Arm Training System

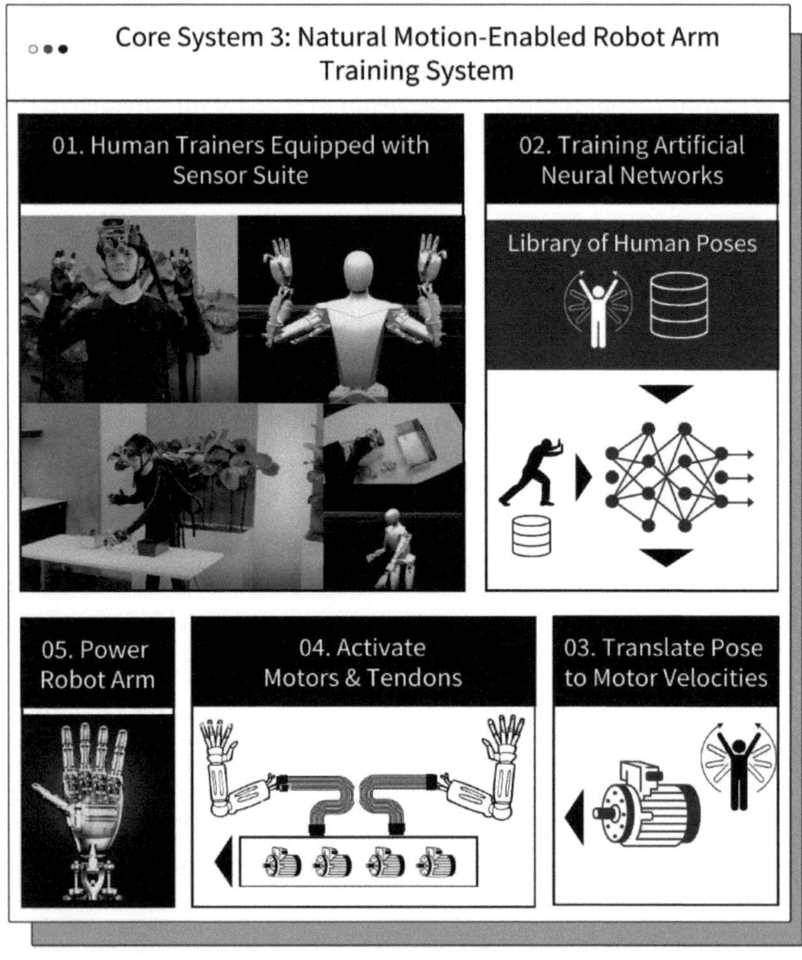

Figure 11.13 — Core System 3: Natural Motion-Enabled Robot Arm Training System

Illustration by the author. Tesla Bot image from Tesla AI Day (2022).

The team building the Tesla humanoid bot employed a group of human trainers and equipped them with an array of sensors, as illustrated in Figure 11.13. Then they provided a list of tasks to the human trainers to perform. As the human trainers performed each of these tasks while wearing the sensor suite, they captured the corresponding motion of their arm, head, torso using the sensor suite and used this dataset to train an artificial neural network, i.e. an artificial intelligence (AI) model. This AI model can now translate any movement that a humanoid has to perform into corresponding motor velocities and drive the robot arm and tendons. Thus, it allows the humanoid to perform a wide range of tasks. It's this artificial neural network that is trained on natural human poses that powers the ultra-dexterous robot arm of Optimus.[25]

Now, there is one major challenge associated with this approach. In theory it works great, but when you deploy a robot trained using this approach in the real-world, you have a 50% chance of success. Why? It doesn't factor into account the wide range of operating environments and object profiles these humanoids will need to handle? True, you can develop a dataset that ensures these human trainers perform the same set of tasks in a range of environments, such as factory floors, homes, warehouses and more. However, that would result in infinite physical training scenarios. Similarly, objects come in potentially infinite shapes and sizes. So how did the team at Tesla address this bottleneck?

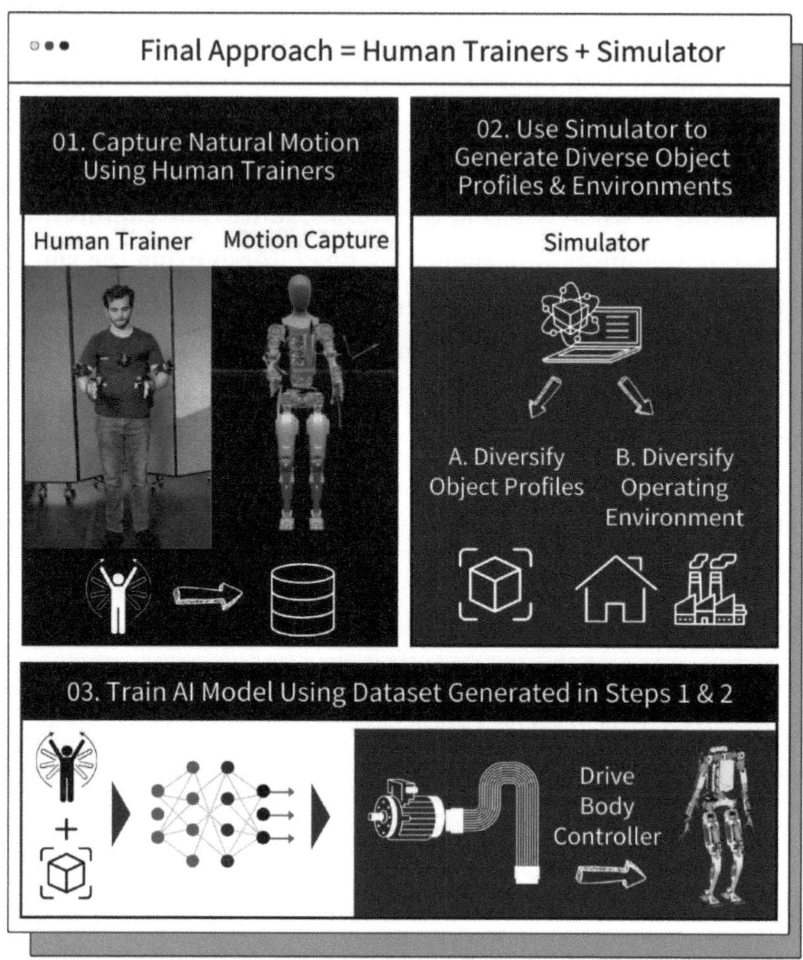

Figure 11.14 — Final Approach = Human Trainers + Simulator

Illustration by the author. Tesla Bot image from Tesla AI Day (2022).

That's where they augmented a second system component to this approach, i.e. simulations, as illustrated in Figure 11.14. The 1st component still remains the same, which means, they still use human trainers to capture the range of motion required for different tasks, but, to handle variables such as the operating environments and object profiles, they use simulators. The simulator allows the team to train this humanoid on 1000s of object profiles and operating environments in a time and resource-efficient manner. This dichotomy of physical training conducted using human trainers, combined with simulations to factor variables, makes this approach extremely scalable, versatile, and robust. In fact, this is the same approach the team adopted to train the body controller. In simple words, this approach was used to train not just the robot arm but basically all its joints to ensure it is able to perform a wide range of tasks with ease and precision.

Now, that naturally leads to the next question: how does this humanoid perceive and navigate this world? That's where the fourth core system comes into picture. We refer to this system as the vision-only autonomous navigation system.

Core System 4 - Vision-Only Autonomous Navigation System

When it comes to enabling autonomy for Optimus, similar to Tesla vehicles, we start with the perception system or the eyes of the robot.

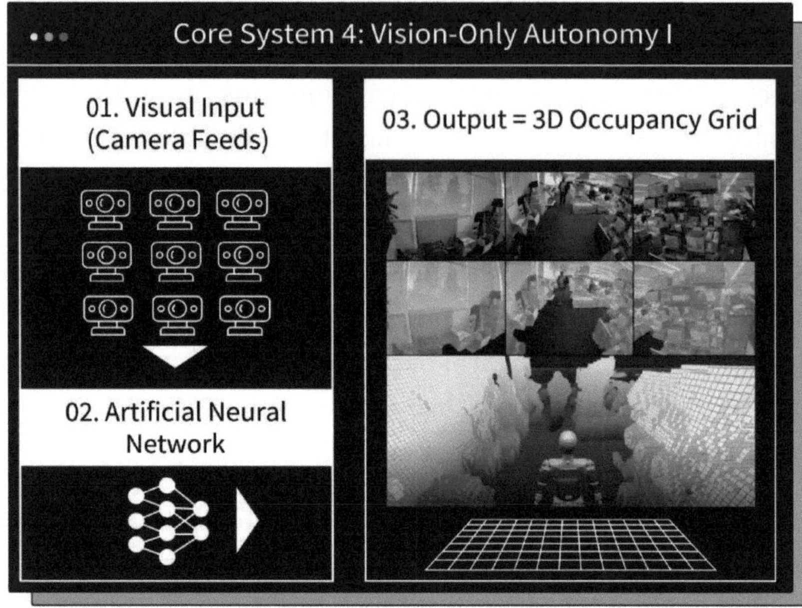

Figure 11.15 — Core System 4: Vision-Only Autonomy I

Illustration by the author. Tesla Bot image from Tesla AI Day (2022).

The team at Tesla has adopted a vision-only approach to develop a perception system for Optimus.[25] This means the Tesla bot is equipped with a range of high-definition cameras that feed continuous streams of visual data to the artificial neural networks running onboard. These artificial neural networks are now trained with indoor datasets since the bot will be operational mostly in indoor environments.

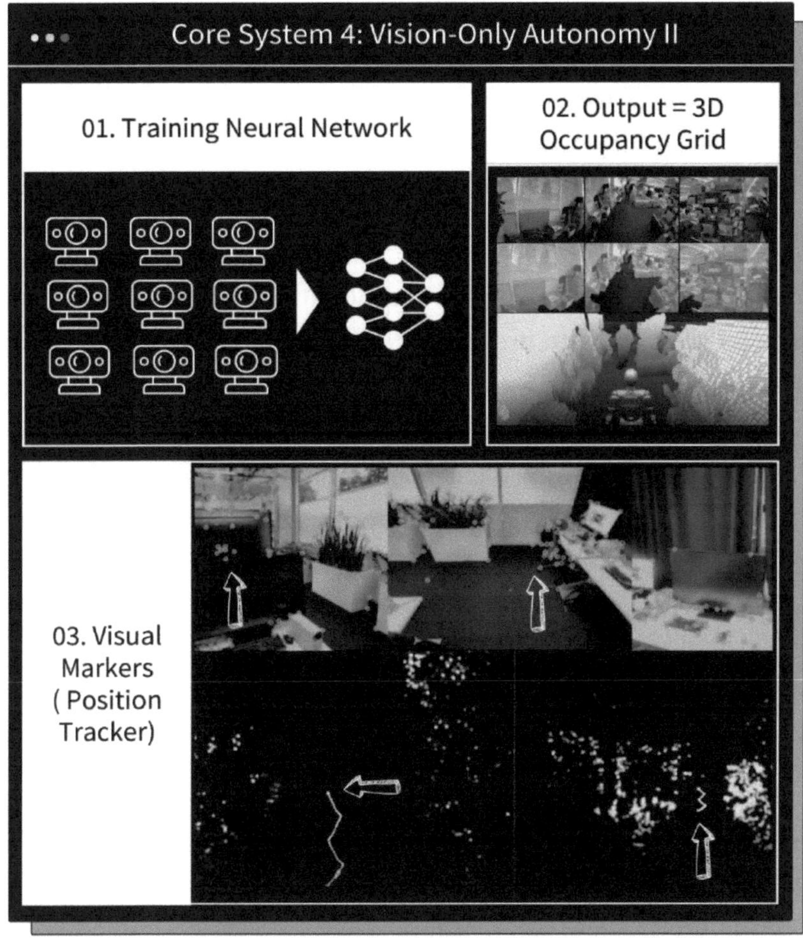

Figure 11.16 — Core System 4: Vision-Only Autonomy II

Illustration by the author. Tesla Bot image from Tesla AI Day (2022).

Once trained, as you can see in Figure 11.16, Optimus develops an occupancy grid which in simple terms means it's able to distinguish free space in front of it and space occupied by obstacles or nearby objects. This is an important aspect of enabling autonomy in humanoids because, without this occupancy grid, it won't be able to develop safe and reliable end to end trajectories for navigation. In simple words, it won't have a clear view of its environment. These occupancy grids are further enriched with visual markers that you see in the bottom half of Figure 11.16. These visual markers correspond to location-specific unique features captured by the camera.

They play a key role especially in helping Optimus identify and track its position with precision. Now that we have the data from the sensors and the artificial neural networks that are being trained to enable autonomy, some of you might ask: Vision alone is not enough, correct? What about balance? How did the team train Optimus to walk in these indoor spaces while retaining its balance? This is where the locomotion planner and simulator comes into play.

Core System 5: Locomotion Planner & Simulator

As you might have guessed from the name, there are two subsystems that are involved here. The first is the planner and second is the simulator. Let's cover each of these one by one.

The goal of the planner is to sense the environment and develop end-to-end trajectories for our robot to navigate safely from source point to destination point, as illustrated in Figure 11.17. So, how does the planner work?

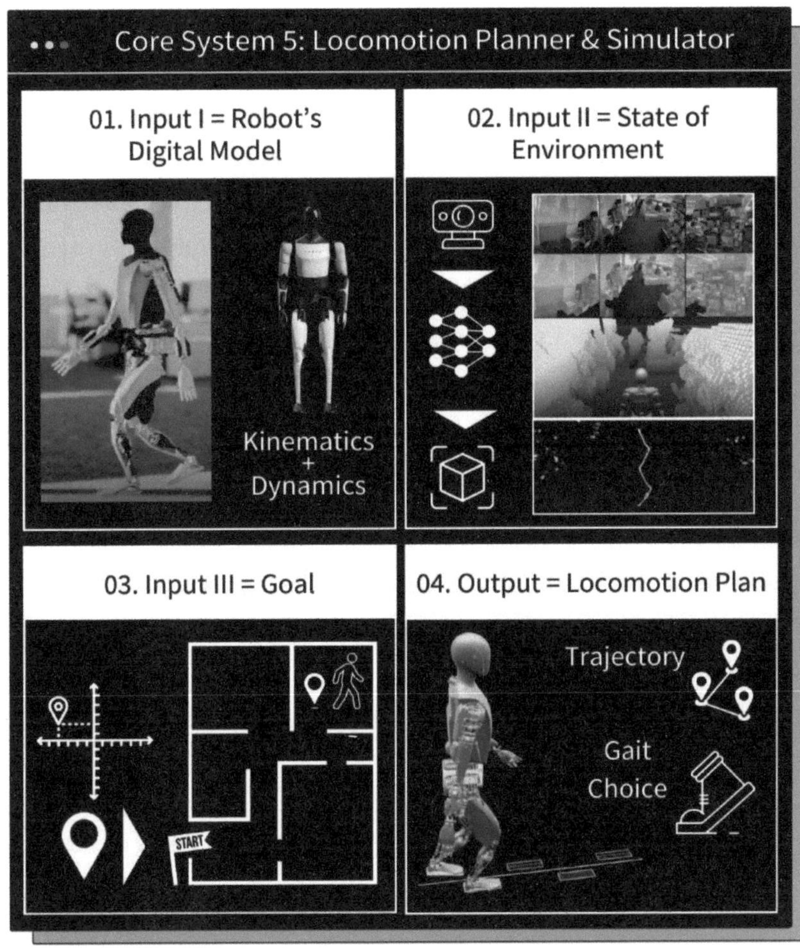

Figure 11.17 — Core System 5: Locomotion Planner & Simulator

Illustration by the author. Tesla Bot image from Tesla AI Day (2022).

First, the team developed a fully detailed digital model of the robot. This model includes Optimus's kinematics and dynamics properties and serves as the first input to the locomotion planner. Then we need to feed this planner the state of the environment, i.e., the perception data. In order for the planner to develop precise trajectories for the robot, it has to have a coherent view of its surroundings, and that's exactly the output of the perception system or the fourth core system. Last but not the least, we need to provide a goal or a destination point to the planner. This is basically the goal point that we want our humanoid to navigate to from its current position. Using this data, the locomotion planner generates an end-to-end route from source to destination point, as illustrated in Figure 11.17. This route, in simple words, translates to a series of waypoints the humanoid must follow to reach its destination safely without running into obstacles and while retaining its balance.

Now, the challenge here with this approach is that it works well in a controlled environment. You may ask: What about the real world? How can we ensure the robot is able to navigate the real world while retaining its balance? That is where the second subsystem comes into play, i.e., the simulator, as illustrated in Figure 11.18.

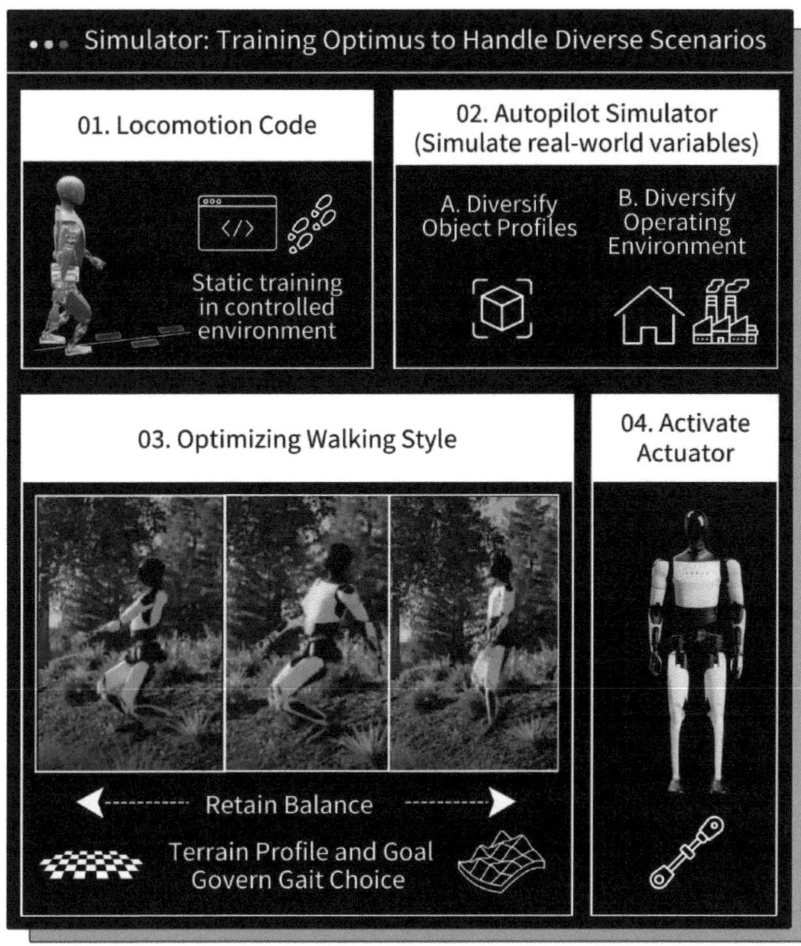

Figure 11.18 — Simulator: Training Optimus to Handle Diverse Scenarios

Illustration by the author. Tesla Bot image from Tesla AI Day (2022).

The team building the simulator for this humanoid started with the Autopilot simulator that was developed for the Tesla autonomous vehicle as a foundational layer. This simulator was repurposed to ensure it can be used to simulate diverse environments, terrain profiles, weather conditions, and other factors that govern the robot's ability to navigate while retaining its balance. More specifically, the inputs from the simulator and the locomotion code now govern the gait choice at each instance of time, or, in simple words they govern the type of footstep the robot chooses at each interval of time. Thus, using this simulator, the team was able to optimize the locomotion code to handle diverse terrain profiles and environmental conditions.[25]

Now, even with the foundational locomotion code and the simulator in place, there is one more challenge that we haven't addressed yet. Any guesses? Did you notice that there is no feedback mechanism in place in this system. That means, this locomotion system that the team has designed up to this point is an open-loop system, and, as with any other open loop system that is static in nature, it will result in high failure rates once it's deployed in the real world. Why?

This static locomotion planner is only able to develop the trajectory for Optimus assuming the world around the robot is more or less ideal and static. That is rarely the case. The reality of our world is that it's constantly evolving. What do we mean by that?

Take any indoor space, whether it's an office space or a factory. The first major variable in the reality space is people. The number of people moving on the factory floor and their trajectories are not necessarily the same each day and at every moment. The same applies to forklifts and other mobile equipment that might be present in the factory. So, there are way more variables in the real world. These variables have the potential to destabilize our bot.

Why? The trajectory or the model generated by the locomotion system is not factoring the real-world conditions into its equation. So how did the team at Tesla solve this problem?

More specifically, how can we collate feedback from the robot's environment as it takes each step and ensure we optimize the choice of gait, i.e., robot's footstep pattern, speed and grip?

That's where the last component comes into play, and we will refer to this as the Motion Control & State Estimation System.

Core System 6: Motion Control & State Estimation System

They developed a high-precision state estimator. In simple words, they refined the locomotion model to ensure it measures reality, i.e., continuously adjusts its output based on real-world conditions. How did they do that?

The team building Optimus used the data from onboard sensors to measure the force that the robot has to sustain as it navigates in the environment. In addition to this sensor data that helps us determine the impact of the environment on the robot's motion, they also used the data from the perception system to better understand what changes in the environment are resulting in this impact. These two streams of data allow the onboard state estimation and motion control system to estimate the exact state of the robot, i.e. its attitude, pose, center of mass and more.[25]

So, as illustrated in Figure 11.19, you can see the attitude and the pelvic pose, which is essentially the vestibular system in us humans, along with the center of mass being tracked when the robot is navigating in the indoor environment.

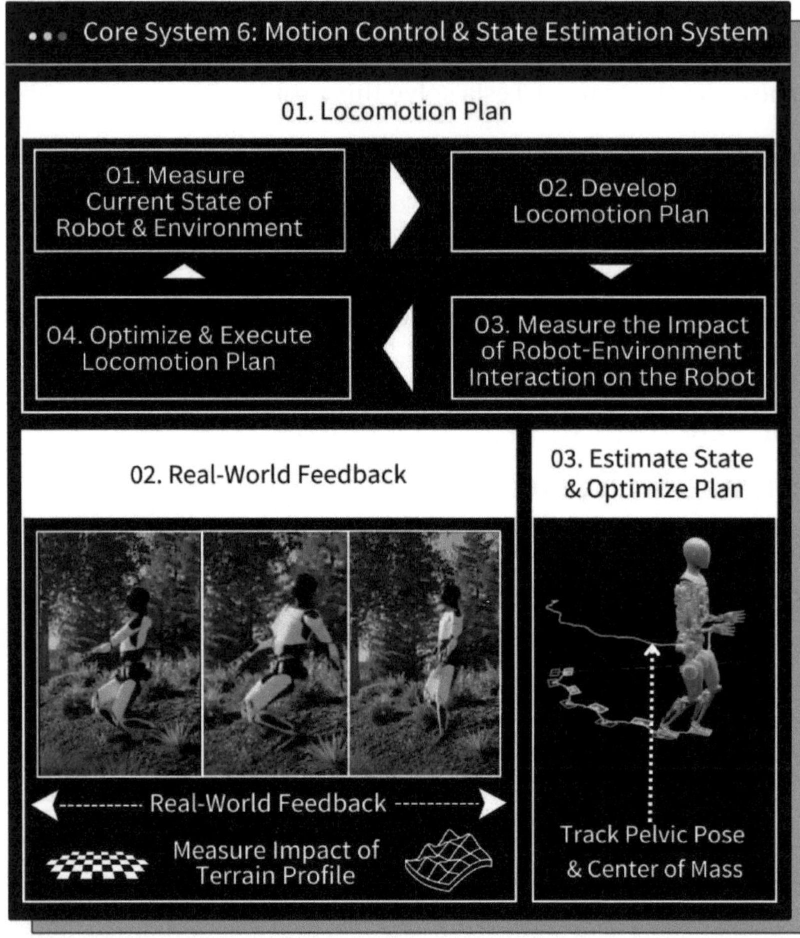

Figure 11.19 — Core System 6: Motion Control & State Estimation System

Illustration by the author. Tesla Bot image from Tesla AI Day (2022).

Now, we have the two key inputs we need. First is the trajectory generated by the locomotion planner, and second is the effect of the real world on those trajectory that the robot experiences, i.e., the attitude and the pelvic post that is being tracked as demonstrated in Figure 11.19. If you think about it, the second input is actually the feedback from the real world for the locomotion planner.

This feedback is important because now our planner is able to feel the world, i.e. can estimate the state of the bot and the state of its surroundings with highest levels of precision. This means our locomotion planner can now calibrate its output or the trajectories to ensure Optimus remains stable, balanced and is able to handle dynamic forces that it encounters as it faces the real world with grace.

That completes the chapter on Tesla's approach to building humanoid robots. While automakers like Tesla are leading the pack on humanoid robotics in the manufacturing sector, the warehouse and logistics industry has not shied away from offering its own blessings to these robots. Companies like Amazon are proof of this trend. In the next few chapters, we will cover how teams at Amazon are building and deploying modern robots at scale.

Chapter 12: Decoding Amazon's Approach to Building Robots

1 million—that's the total number of robots deployed by Amazon across more than 175 fulfillment centers around the globe. Each fulfillment center is a massive warehouse, averaging about 1 million sq. ft., which is roughly the size of 20 American football fields. Each warehouse is powered by 9 types of robots that help employees fulfill customer orders. From sorting and packaging to lifting shelves and more, these robots assist employees with a wide range of tasks.[26-27]

Amazon processes around 6 billion orders each year, or about 17.2 million deliveries per day.[28] More than 75% of those orders are now handled by one of these 1 million robots. Given this scale of adoption, it has become more important than ever to understand how Amazon is using modern robots to scale its operations around the globe.

Amazon’s Journey in Robotics: 1 Million Robots

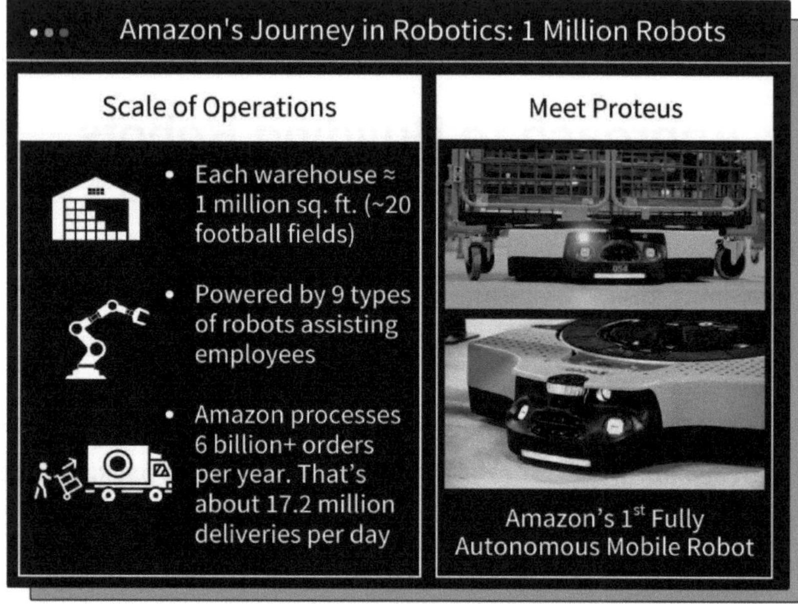

Figure 12.1 — Amazon's Journey in Robotics: 1 Million Robots

Illustration by the author. Amazon Proteus robot image from About Amazon (2024).

To begin with, Amazon's journey in robotics started around 2012, when the company acquired Kiva, a robotics startup building autonomous mobile robots capable of shifting shelves full of packages around the warehouse.[9] Amazon immediately integrated this fleet of Kiva robots across its warehouses, and after seeing double-digit gains in operational efficiency, it became clear that if it was to continue fulfilling more than 6 billion orders each year, or about a million each day, robots were the only way forward.

Since then, it has integrated nine types of robots into its global warehouses. Each of the nine robots developed by Amazon is custom-designed to perform specific tasks such as packaging, sorting, tagging, and more. What started as an autonomous shelf transit project in its warehouses with the adoption of Kiva mobile robots has now resulted in the development of next-generation fulfillment centers equipped with a wide of inventory robotic arms like Robin, Cardinal, Sparrow, and Sequoia and autonomous mobile robots like Proteus, Hercules, and Titan.

Amazon refers to Sequoia, as "a state-of-the-art multilevel robotics-enabled containerized inventory system that makes it faster and safer for employees to store and pick goods." The Sequoia robotic system can handle more than 30 million items on any given day. This scale of adoption naturally raises the question: How does Amazon's robot-enabled warehouse automation system work end to end?

Inside Amazon: A Package's Journey Through 9 Robots

To begin with, it's important to remember that there are more than 175 fulfillment centers located around the world. These, in simple terms, are essentially core warehouses stacked with products that are in demand in these 175 regions. The process starts with Amazon's inventory planning system, which ensures that the products most likely to be in demand in each region are stocked in the corresponding fulfillment centers near customers. This inventory planning system is trained using artificial neural networks on the database of customer orders received each day.

Once the company makes sure that there is enough inventory in each warehouse, the next step is to consolidate all orders received each day and transfer the corresponding items from the inventory area to the employee workstation, so employees can pick out items and assist with packaging and sorting. This transfer of items from the inventory section of the warehouse to the employee workstation is carried out using a combination of industrial robotic arms and autonomous mobile robots, such as those originally introduced through Amazon's acquisition of Kiva Systems. Together, these robotic systems orchestrate the internal flow of goods across the warehouse, as illustrated in Figures 12.2 and 12.3.

It is these nine types of robots that enable Amazon to fulfil approximately 17.2 million orders per day.[27] So, what are these nine robots, and how do they work?

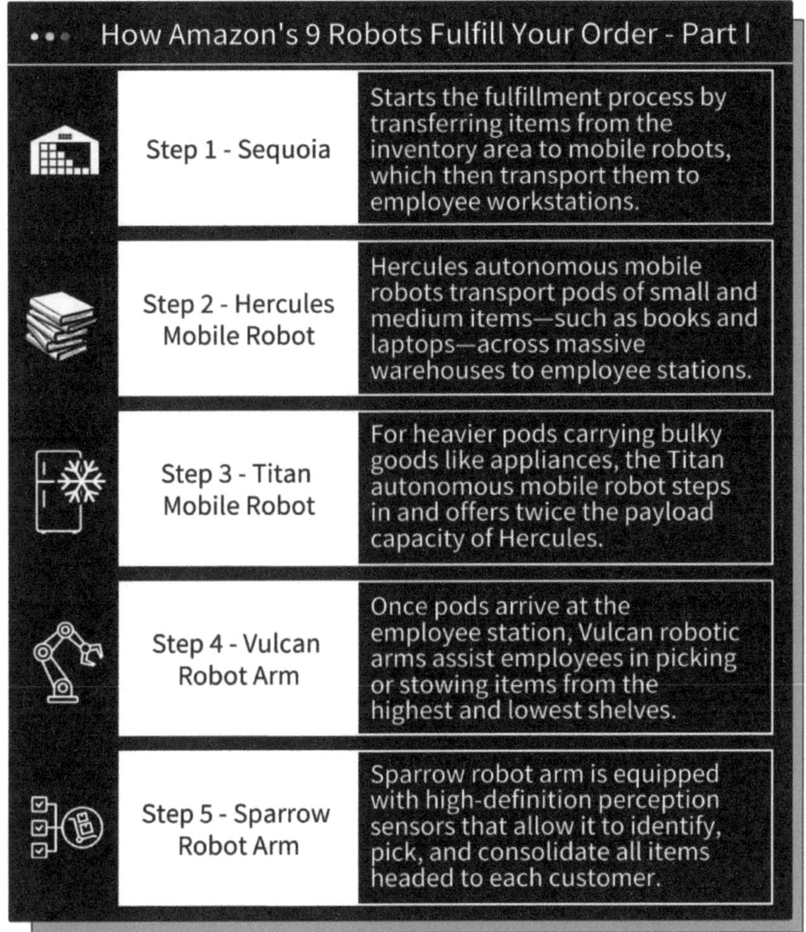

Figure 12.2 — How Amazon's 9 Robots Fulfill Your Order - Part I

Illustration by the author.

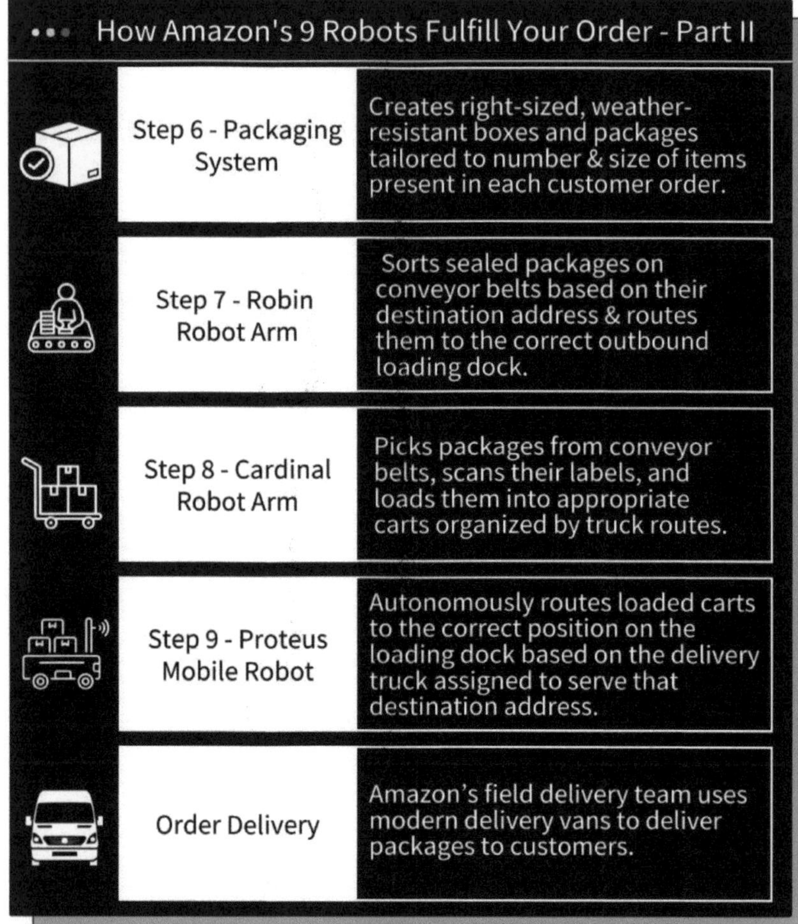

Figure 12.3 — How Amazon's 9 Robots Fulfill Your Order - **Part II**

Illustration by the author.

Robot 01: Sequoia

The process begins with Sequoia, a giant robotic arm-enabled inventory-management system that is installed in each warehouse. This system is responsible for consolidating all the items that need to be packaged and shipped per day from the inventory area of the warehouse to the mobile robots. Once these items are consolidated, they are transferred to a series of pods and loaded onto autonomous mobile robots called Hercules and Titan.

Robot 02: Hercules

Next, Hercules autonomous mobile robots (AMRs) transport pods filled with small and medium-sized items like books, bottles, computers, and more to employee workstations. Hercules navigates the warehouse autonomously by following a grid of encoded markers printed on the warehouse floor. It reads these markers using a forward-facing 3D camera, which helps it to identify both its own position and surrounding obstacles and make safe routing decisions. Each of these Hercules robots is capable of lighting up to 1,250 pounds and can cover over 1 million sq. ft. per warehouse. Each weighs about 300 pounds.[27]

Now, some of you might ask: What about the heavier goods? How does Amazon transport beds, mattresses, and other heavy goods across its warehouse? That is where Titan, Amazon's other autonomous mobile robot, comes into play.

Robot 03: Titan

For heavier goods like appliances or furniture, Titan (AMR) steps in with twice the payload capacity of Hercules. It ensures that even the bulkiest pods, such as those carrying refrigerators or similar large-sized items, are transported efficiently across the warehouse. While Hercules and Titan enable the transfer of pods from the inventory area to the employee workstation, it's important to remember that these pods are humongous. This means employees may need to use a ladder to climb to the top of a pod to fetch an item or, worse, bend significantly if they need to retrieve something from the bottom of the pod. So how does Amazon solve this problem? That is where the Vulcan Robot Arm comes into the picture.

Robot 04: Vulcan

Once pods arrive at the workstations, employees use the Vulcan robot arm to collect items stored at the very top or bottom of the pod.

There are two versions of Vulcan: Vulcan Pick and Vulcan Stow, both deployed at Amazon fulfillment centers. Vulcan Pick, as the name suggests, picks items located at the top rows of the inventory pods. It can grab items up to 5 pounds and 14 inches in length. Launched recently at a fulfillment center in Hamburg, Germany, the Vulcan system can reach as high as 9 feet from the ground.[29] Similarly, Vulcan Stow is another robot arm that is used to stow items back into the inventory area.

What's important to remember is that it needs to handle both fragile items like ceramic cups and LCD displays, as well as sturdier ones like books and containers. So how does it do that?

The team building Vulcan designed it with these requirements in mind, while also ensuring an ergonomically safer solution that helps employees access hard-to-reach areas without straining their back or spine. To meet these requirements, the Vulcan robot arm is equipped with three core components: a camera, a suction cup, and a force measurement sensor. The camera allows it to detect the right objects to pick up, the force measurement sensor ensures it applies just the right amount of force based on the object profile detected using the camera and the underlying AI model, and the suction cup allows it to securely grasp the object. Vulcan works in tandem with employees to ensure higher operational efficiency.

Robot 05: Sparrow

Now that items have been transferred from the pod and placed on employee workstations, the next step is to consolidate all items headed to each customer. In other words, all items for a given order need to be placed into a single tote or basket so they can move on to the packing stage. How does Amazon solve this problem?

That's where the Sparrow robotic arms come in. Sparrow can lift packages of up to 12 pounds and is about 7.5 ft. long, roughly the size of an elephant's trunk. Powered by AI-enabled cameras and sensors, Sparrow can accurately identify all items headed to each customer and place them into the correct totes, ensuring every order is consolidated into one customer-specific basket.

Robot 06: Packaging System

These totes, or customer-specific baskets, then move to the automated packaging system. Now, each customer order includes items that come in different shapes and sizes. This means we need a robotic system in place to detect the type of order, the items present in the order, and curate a compatible-sized package that will fit all the items in the order. This is an extremely tedious process, and Amazon handles this step using a series of robotic packaging systems.

This robot-arm-enabled packaging system handles about thousands of customer orders per week and custom-manufactures packages to fit an order's dimensions, thereby eliminating material wastage. The system uses a weather-resistant paper composite as raw material and outputs protective bags and boxes sized precisely for each order. This step reduces material waste while ensuring every package is protected.

Now that the boxes are ready to ship, they need to be sent to different zip codes based on the corresponding destination address of the user who ordered those items. That means these packages now need to be routed to appropriate delivery trucks, each covering a specific geographical region under the delivery network of this warehouse. So how do we make sure this massive flow of packages is sorted by destination and routed to the right outbound trucks?

That’s where the next robot, Robin, comes into the picture.

Robot 07: Robin

Once boxed, Robin robotic arms sort packages by destination, transporting them across a series of conveyor belts to the correct outbound docks. Robin can lift packages of up to 50 pounds and was the first robot arm ever deployed by the Amazon Robotics team. It is about 6 ft. long, roughly the size of an elephant’s trunk. This process of routing packages across a series of conveyor belts located in the warehouse using the Robin Robot Arm ensures every package is routed to the right region for delivery.

These packages then end up in the outbound dock area, the official endpoint of Amazon's fulfillment center. Here, they are lifted from the conveyor belt and placed into carts, which are later transported to the delivery truck. To enable this process of transferring packages from belt to cart, Amazon designed another robot arm called Cardinal. So, how does Cardinal work?

Robot 08: Cardinal

At the outbound docks, Cardinal robotic arms take over. These robot arms lift packages from the conveyor belt using an air suction cup, read the label, and precisely place each package into the appropriate cart before it is brought to a truck in the loading dock. In simple terms, packages need to be routed to the right trucks based on their labels, i.e., destination address and service, and this job is handled by the Cardinal Robot Arm.

Cardinal can lift up to 50 pounds and load hundreds of packages per hour. Now that the packages are sorted into outbound carts, there is still the task of transporting these carts to the truck. So, how does Amazon handle this process? This is where the ninth robot, also known as Proteus, comes into the picture.

Robot 09: Proteus

Proteus autonomous mobile robots (AMRs) route the loaded carts filled with outbound packages to appropriate positions on the loading dock. From there, the packages are transferred to the truck.

Proteus is the first fully autonomous mobile robot developed by Amazon, meaning it can route freely across the warehouse using a suite of sensors and navigate safely around obstacles.[30] You may ask: What about other mobile robots like Titan and Hercules that we discussed earlier? Those are confined to dedicated robot-compatible areas within the warehouse marked with floor markers. Thus, Proteus and Cardinal, therefore, work in tandem to ensure packages are routed to the loading dock and made ready for last-mile delivery. So, given Amazon needs to ship more than 17 million packages per day, how many packages can be transported into a single truck?

Each Amazon truck carries about 2,000 packages.[31] These packages are transported to a variety of locations, including customer sites, Amazon Air sites, and more. Amazon Air sites are facilities where packages are transferred into containers that are loaded onto one of the 110 aircraft owned by Amazon. Once an aircraft lands at its destination, the packages are moved to another fulfillment center, where the same process starts again before the order is delivered to the customer.

Since 2012, over the last decade, Amazon has significantly ramped up its warehouse automation initiative. Recently, it deployed its one millionth robot at a fulfillment center in Japan and is currently training a humanoid at an indoor facility in San Francisco. That leads to a question: With 1 million+ robots deployed across 175+ warehouses, how does Amazon ensure these robots do not collide with each other? More importantly, how do you route robot-led traffic at this scale? We will cover that in the next chapter.

Chapter 13: Amazon's One-Million-Robot Traffic-Jam Problem

6 billion orders fulfilled across 25 countries by one million robots working in collaboration with 1.5 million employees: that's the story of Amazon. In the previous chapter, we learned how Amazon kickstarted its journey in robotics through the acquisition of Kiva Robotic Systems and subsequently developed 9 custom robots that now power its 175+ warehouses. To be specific, it has replicated these 9 robot types and built multiple fleets of robots for every warehouse. totalling a staggering one million robots throughout its global operations.[27]

While adding robots to its warehouses has significantly elevated its operational efficiency, this very addition has also given rise to a series of techno-operational bottlenecks. From enabling real-time communication between robots and humans, to addressing massive robot traffic jams in these warehouses, and training humanoids to scale last-mile deliveries, there are a series of challenges that the Amazon robotics team is now trying to address with the help of modern AI models.

After all, how do you build a human–robot collaboration system that spans from the warehouse to your home and helps you automate each leg of this journey at this scale? This system must work across a fleet of one million robots and a workforce of 1.5

million employees.[32] That is the question we will try to address in this chapter. To begin with, it helps to first understand the challenge of robot-led traffic jams in these warehouses.

Amazon's Challenge: 1M Robots, 175+ Warehouses, 1 Giant Traffic Jam.

If we zoom out a bit here and try to visualize the scale of operations at Amazon, this is what that picture might look like. Imagine, 175 warehouses covering 25 countries, totalling one million robots.[27] That math boils down to about 5,700 robots per warehouse. Now it's important to remember here that these 5,700 robots come in nine forms. We covered these robots in detail in the previous chapter. Some of these are autonomous mobile robots, and some are robotic arms. How is Amazon managing a fleet of one million robots that come in nine forms and are deployed across 175+ warehouses, each with its own layout and traffic patterns? Figure 13.1 illustrates the scale of this challenge in detail.

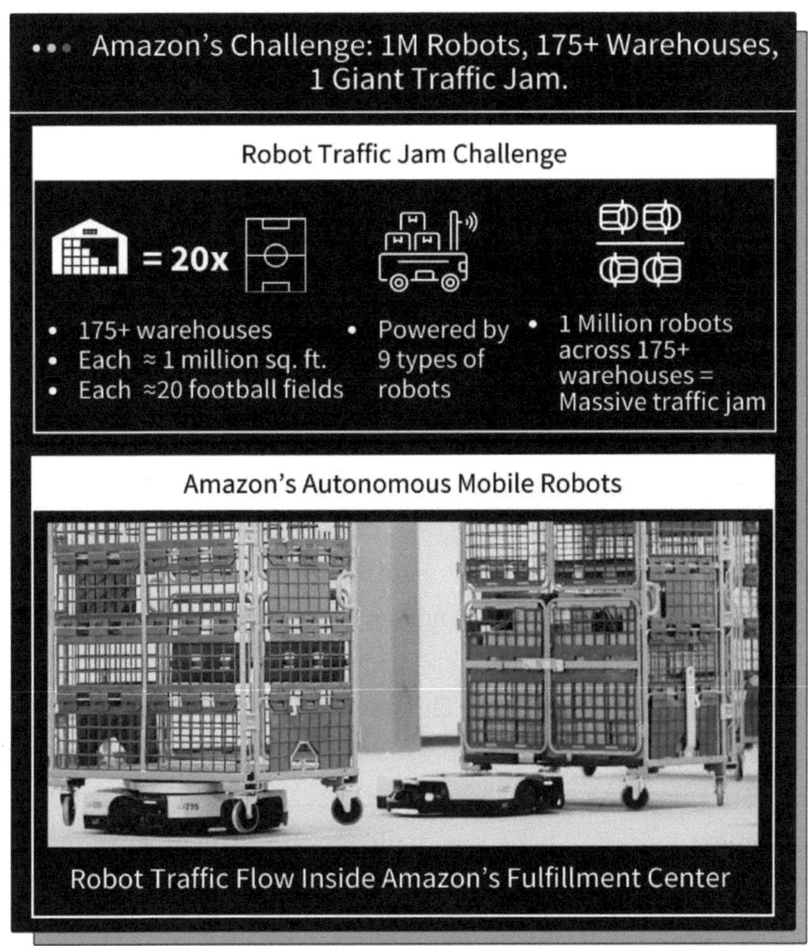

Figure 13.1 — Amazon's Challenge: 1M Robots, 175+ Warehouses, 1 Giant Traffic Jam

Illustration by the author. Amazon Proteus robot image from "Meet Amazon's First Fully Autonomous Mobile Robot," Amazon News YouTube channel (2022).

To simplify that question, let's try to do the rough math and understand this challenge for a single warehouse, and then we will talk about all the warehouses. Today, there are about 5,700 of these robots deployed in each of its warehouses around the globe. For simplicity, let's assume half of them are mobile robots and the remaining half are static robot arms. That leads us to roughly 2,850 mobile robots per warehouse. Let me remind you: each of these warehouses averages about one million sq. ft. So now, imagine being a robotics engineer trying to solve the traffic jam in a one million sq. ft. space for about 2,850 robots. Can you pause here and just imagine the level of challenge? We are trying to develop a traffic management solution for roughly 3,000 robots packed into a one million sq. ft. warehouse.

Now, as a robotics engineer, try to add on top of that the realities of running robots at this scale in a production environment. What does that mean? Robots will need to be recharged, Wi-Fi might drop unexpectedly, an employee might unknowingly step into robot-only designated zones, electricity outages may occur, robots may malfunction, and more. This means we cannot have a static routing solution for robots in these warehouses. We cannot simply program rules for each of these robots because the edge cases here are almost infinite. That's the challenge the team at Amazon ran into. So, how did they solve it?

They built DeepFleet, an AI-enabled traffic-management model that predicts and prevents robot-led congestion in real time across its warehouses.[33] Some of you might ask: why not stick with simulations? Why use an AI model? Isn't that the solution that most robotics companies have been using for a long time?

Why Not Just Simulate?

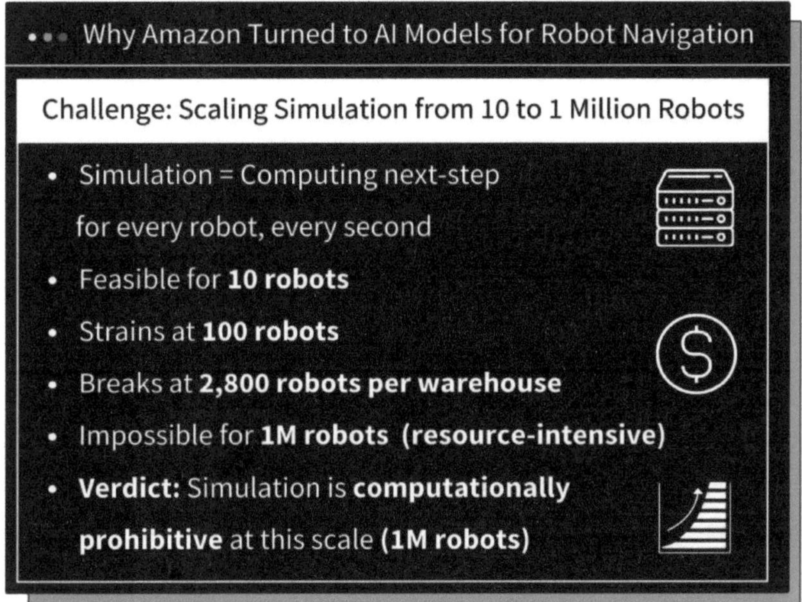

Figure 13.2 — Why Amazon Turned to AI Models for Robot Navigation

Illustration by the author. Amazon Proteus robot image from "Meet Amazon's First Fully Autonomous Mobile Robot," Amazon News YouTube channel (2022).

Simulation really boils down to computing the next step of each robot using the existing robot navigation algorithm, for all robots in the warehouse, for every second in the future. Can we do that for 10 robots? Sure, we can. What about 100 robots? Maybe. What about 2,800 robots? Theoretically, it's possible, but in reality, we don't have enough computing power to run simulations for 2,800 robots per warehouse, totaling one million robots.

So, while simulation could technically yield highly accurate results, in practice it consumes enormous computing power. This makes it a highly resource- and time-inefficient solution if you're planning to run it for 2,800+ robots in a single warehouse, which is the case for Amazon for each of its warehouses. That's why Amazon instead built a foundational AI model to enable robot navigation. That naturally leads to the question: so how does an AI model fundamentally solve this problem?

Why build a Foundational Model?

An AI model, at the end of the day, is a giant probability equation with thousands of variables trained on petabytes of data. Its real strength lies in recognizing patterns. Instead of trying to compute the next move for every robot in parallel at every second, it identifies and filters traffic patterns from the training data and uses those patterns to predict where congestion is likely to occur in the warehouse. Here, we are no longer computing the next move for all robots for every t+1 second; rather, we are making predictions or estimations based on historical data, and this prediction, or the approximation function, reduces the number of computations we will need to perform and, hence results in a time- and resource-efficient solution.

In other words, this AI model is trained to extract traffic patterns from historical (training) data and apply them holistically across the warehouse floor (test data), forecasting congestion points before they emerge. Put simply, it's pre-trained to recognize patterns and continuously learns as new patterns emerge. This approach greatly reduces the resources needed for computation and makes the process far more time- and resource-efficient.

So, how did Amazon develop the DeepFleet AI model to actually solve the traffic-jam problem in the warehouse?

DeepFleet: Amazon's AI Model for 1 Million Robots

Before we try to understand how it developed this foundational AI model, it is important to remember that when Amazon refers to DeepFleet, it is referring to a suite of pre-trained AI models that helps the team analyse the human–robot–goods traffic flow across the warehouse floor from different perspectives, and work together to prevent 2,800+ robots from bumping into one another.[33]

What does a human–robot–goods flow mean in this context? Well, if you take a look at the warehouse operations, there are human workers on the floor, there are robots, and there are goods as well as goods-carrying machines. The team factored all these variables into its equation as it began building this suite of AI models. That naturally raises the question: how did they even build this suite of AI models?

Instead of hand-coding routes for each robot or running compute-heavy simulations, the team building Amazon robots built a total of 4 AI models by training artificial neural networks on billions of hours of navigation data captured from about one million robots operating in its 175+ warehouses around the globe, as illustrated in Figure 13.3. The goal: predict congestion before it happens and reroute robots in real time in each of these warehouses.

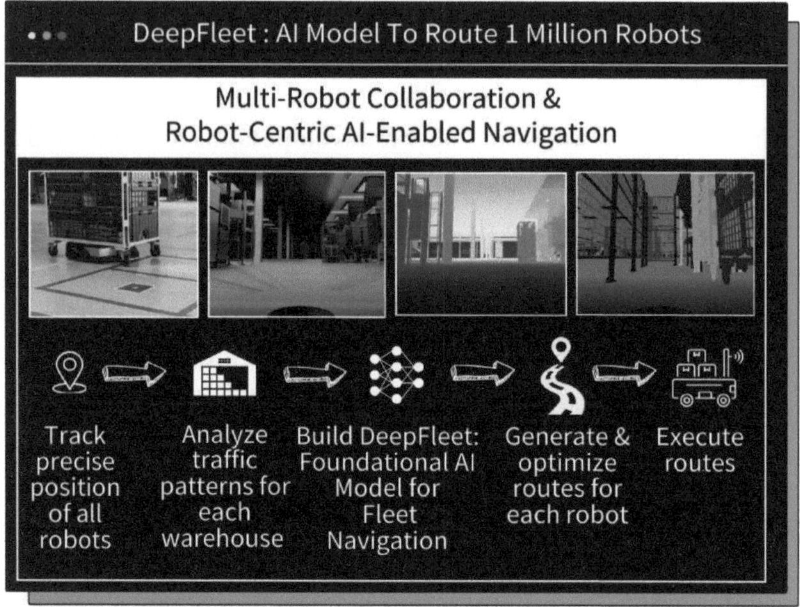

Figure 13.3 — DeepFleet: AI Model to Route 1 Million Robots

Illustration by the author. Amazon Proteus robot image from "Meet Amazon's First Fully Autonomous Mobile Robot," Amazon News YouTube channel (2022).

Once it developed these models, it evaluated each of them across its warehouses and concluded that certain models were more suited for accurately predicting the next move of the robot, and others were more suited for holistically looking at all the robots on the floor and predicting overall congestion patterns. Similarly, from a resource-efficiency perspective, certain models were resource-heavy and others were resource-efficient.

So, how did it decide which model to use when, and, more importantly, what are these models?

The Four AI Models That Make Up DeepFleet

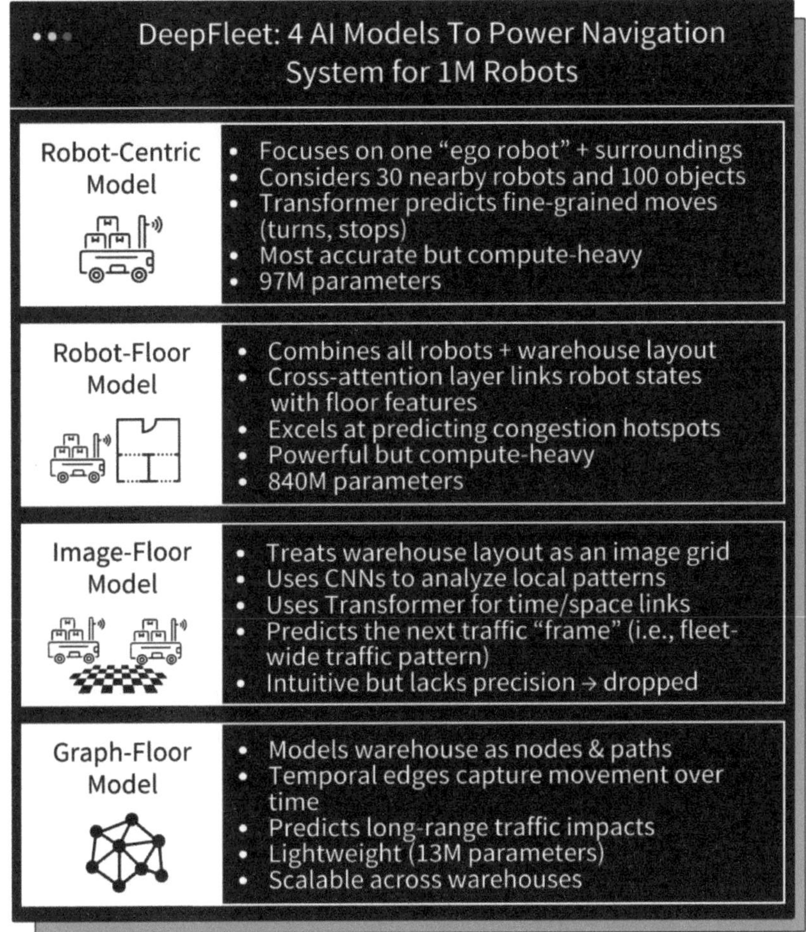

Figure 13.4 — DeepFleet: 4 AI Models To Power Navigation System for 1M Robots

Illustration by the author.

To build DeepFleet, Amazon experimented with four distinct models, each offering a unique "lens" on the warehouse, as illustrated in Figure 13.4. We will now cover each of these one by one.

Robot-Centric Model – One Robot's View of the Warehouse

At a high level, the robot-centric model looks at one robot at a time, called the 'ego robot', and its immediate surroundings. It considers where the robot is, where it's headed, and how nearby robots and objects might influence its movement. By repeating this process for every robot on the floor, Amazon can predict the fleet's collective movement. This makes the model highly accurate at analyzing traffic patterns at local level, or robot level, making it ideal for forecasting fine-grained actions for each robot, like turns, stops, and lane changes.

On the technical side, the model processes far more information than just the state of the ego robot itself. It encodes the state of the ego robot alongside the 30 closest robots, 100 nearby grid cells, and 100 surrounding objects like pods, chutes, and charging stations. A Transformer then pulls this information together to predict the robot's next step. When scaled to a warehouse with 2,800 robots, this approach delivered the highest accuracy of all the models Amazon tested. The trade-off is size: with 97 million parameters, it requires significant computing resources to run, but the surgical precision it provides makes it a critical piece of DeepFleet.[33]

While this model is heavily centered on the ego robot, or one robot at a time, some of you might ask: what about the overall fleet pattern? That is the job of the next model.

Robot-Floor Model – Factors All Robots & Global Floor Map

The robot-floor model shifts perspective from the individual robot to the fleet. Instead of focusing on one robot at a time, it combines the state of all robots in the warehouse along with the overall floor layout into a single equation. This means each robot's decision is guided not only by what its immediate neighbors are doing, but also by the broader traffic patterns emerging across the warehouse. It factors in the global warehouse context—lanes, pods, obstacles, and drop-off points—along with the simultaneous movements of all robots on the floor. As a result, the robot-floor model excels at anticipating when and where congestion is likely to appear, particularly at chokepoints such as intersections or busy transfer zones.

Technically, the model encodes the aggregated state of all robots on the floor and floor features separately, then merges them through cross-attention layers. This allows the system to reason about both robot-to-robot interactions and robot-to-floor relationships simultaneously. The approach proved highly effective in predicting the timing of congestion events, giving planners a critical edge in managing traffic in these warehouses.

The trade-off is scale: with 840 million parameters, it is by far the largest and most resource-intensive of the four models, making it powerful but expensive to run.[33] This naturally raises the next question: how did Amazon manage the resource-intensive demands of such a model? That is where the next model comes in.

Image-Floor Model – Warehouse as an Image Capturing Traffic Flow

The image-floor model takes a more visual approach. It treats the warehouse floor as if it were an image, with each grid cell acting like a pixel that encodes whether it contains a robot, an obstacle, or is unoccupied. By scanning this 'image' with a convolutional neural network (CNN) and then passing it through a Transformer layer, the model attempts to predict the next 'frame' of the warehouse traffic pattern. The convolution neural network here excels at identifying features in images and, hence, allows us to detect local spatial patterns: clusters of robots, blocked lanes in different regions of the warehouse. The Transformer then uses its attention mechanism to connect those local features across both space and time. In simple words, it can relate a group of robots in one corner with another group far across the warehouse, or it can link today's congestion hotspot to how traffic evolved in previous seconds.

To simplify, CNNs allow us to identify local patterns around the warehouse and the Transformer connects these patterns over time and gives us a global perspective that spans space and time. In concept, this approach is intuitive—just as video models can predict the next frame of a movie, the image-floor model tries to predict the next frame for the robot traffic pattern in the warehouse.

In practice, however, the results were underwhelming. This model struggled to deliver the precision needed to accurately forecast robot traffic patterns. As a result, the image-floor model underperformed compared to other models and was eventually set aside.[33] While it offered a creative lens for thinking about the problem, it lacked the accuracy and reliability required to support Amazon's real-world operations. That led to the development of the next model, i.e., the graph-floor model.

Graph-Floor Model – Warehouse as a Map of Nodes & Paths

To understand the graph-floor model, it helps to first know what a graph system is. In the context of computer science and mathematics, a graph system is made up of nodes (points) and edges (lines that connect those points). A good everyday example is a subway map: the stations are nodes, and the tracks between them are edges. Graphs are powerful because they allow us to visualize the connections between different datapoints present in our dataset.

Amazon used this idea to reimagine the warehouse floor. Each floor cell is treated as a node, and the valid moves between cells—say, moving north, south, east, or west—are treated as edges. In effect, the warehouse becomes a giant connected map of paths a robot can take inside the warehouse, just like a subway map. To capture the passage of time, Amazon also added temporal edges—extra connections that link the same cell to itself at the next time step. This allows the system to reason not only about where robots are, but also about how their positions will change in the near future. With this setup, the model is especially good at predicting long-range interactions, such as how a traffic jam in one corner of the warehouse might impact the other areas.

Despite being much smaller than the other models, only about 13 million parameters, the graph-floor model delivered relatively strong overall performance.[33] While it didn't reach the same peak accuracy levels as the robot-centric or robot-floor models, its lightweight design allows it to scale across Amazon's 175+ warehouses, each with unique layouts and traffic patterns, making it a valuable tool for large-scale deployment.

Now that we have covered all four models, the question becomes: how did they perform when the Amazon team evaluated them against one another? And what, ultimately, was the outcome of this evaluation process?

How the Models Performed and Are Used Today

When Amazon evaluated the four models, the robot-centric model came out on top for accuracy, making it ideal for fine-grained, per-robot predictions such as turns, stops, and lane changes. The robot-floor model excelled at forecasting congestion timing and locations, but its massive size—840 million parameters—made it extremely expensive to run. The image-floor model, while conceptually appealing, consistently underperformed and was set aside. The graph-floor model proved less precise than the first two, but its efficiency and scalability made it valuable for long-range traffic prediction across Amazon's diverse warehouse layouts.

Today, Amazon relies on three of the four: robot-centric for local precision, robot-floor for system-wide congestion awareness, and graph-floor for efficiency at scale. Together, these models give DeepFleet a layered view of its operations, capturing individual robot moves and global traffic patterns, enabling Amazon to coordinate over one million robots across its 175+ facilities.

Beyond addressing robot traffic jams, there was another bottleneck they had to address. Any guesses? Building and routing one million robots is one aspect of this equation, but let's not forget these robots are working in tandem with humans, i.e., warehouse staff. So how do you enable interaction between these two agents?

More specifically, we are referring to about 1.5 million employees and 1 million robots. So how do you develop a human-robot interaction channel at that scale? Another question on those lines: so far, we have only covered the role of robots inside the warehouse; what about outside the warehouse? What about last-mile deliveries? We will cover these questions in the next chapter.

Chapter 14: Decoding Amazon's Human–Robot Interaction Systems

Today, Amazon operates about 175+ warehouses, each spanning 1M+ sq. ft and is packed with 1,000+ robots working in tandem with 1,000+ employees.[26] This is human-robot collaboration at galactic scale. How do you even think about building a solution that allows these two species to communicate with each other in real time? How do you ensure that 1.5 million employees can communicate and work with these 1 million+ robots in real time, while keeping the process hands-free and time-efficient? We will try to address this problem and the corresponding solution in detail in this chapter.

Voice-Controlled & AI-Agent-Enabled Warehouse Robots

At each warehouse, robots and humans need to work in tandem, which means they must be able to communicate with each other without disturbing the operational efficiency of the warehouse. What does operational efficiency mean in this context? Let's say a human assistant working at the facility needs a robot to focus on a specific task or pause for a few minutes to enable repair work in that space. The human worker must be able to communicate that to the robot in a time-efficient and hands-free manner. So, how do you solve this problem?

Some of you might ask: what if we provided a joystick to each employee? A joystick may not be a scalable solution here because we are talking about thousands of workers at each facility interacting with nine fleets of robots. So, how did Amazon solve this problem? What was their approach?

To address this challenge head-on, Amazon is integrating voice-enabled AI agents into its operations. This means employees can interact with robots directly through voice assistants.[34] The approach adopted to build these agents is illustrated in Figure 14.1.

This solution results in two major effects on its operations. The first-order effect here is that employees can now communicate directly with the robots hands-free and in a time-efficient manner, using natural language. So, it not only solves the communication problem but also minimizes the training time needed per employee post onboarding. That's the first-order effect.

The second-order effect is that we are now transitioning from the old world of training robots to perform singular tasks to enabling them to handle a variety of tasks. How? Robots are now capable of understanding commands received in natural language, which means they can be trained to take each of these commands and perform the corresponding tasks. This shift marks the beginning of Amazon's entry into the world of next-generation human-robot interaction systems. It represents a significant change from the last decade, as Amazon now focuses on building its next generation of robots—humanoids capable of performing a wide range of tasks delegated through voice commands by warehouse managers.

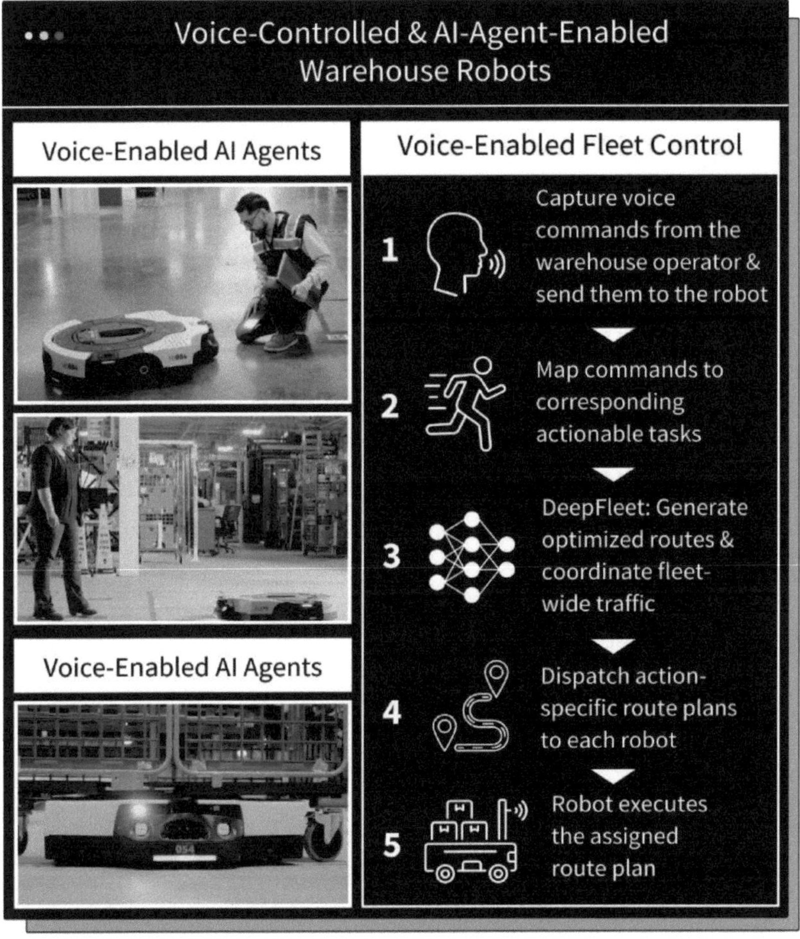

Figure 14.1 — Voice-Controlled & AI-Agent-Enabled Warehouse Robots

Illustration by the author. Amazon robot image from About Amazon (2023)

Some of you might ask: These efforts only address package fulfillment bottlenecks within the warehouse. What about package delivery? How is Amazon planning to use robots to automate deliveries outside the warehouse?

Training Humanoids in San-Francisco Lab for Package Deliveries

To automate last-mile deliveries, Amazon is planning to equip its current fleet of 20,000 Rivian electric vans with humanoids. It is currently testing humanoids with one of these vans at its newly built 'humanoid park,' an indoor facility in San Francisco custom-designed to serve as a training ground for its next generation of humanoids. While Amazon has yet to reveal its detailed plan for humanoids, Digit—a humanoid developed by Agility Robotics, one of Amazon's key partners—appears to be the first candidate in this evaluation process.[18,35]

The approach adopted by the team at Amazon to train Digit is illustrated in Figure 14.2.

Figure 14.2 — Training Humanoids to Scale Amazon's Last-Mile Delivery Network

Illustration by the author. Digit humanoid robot image from About Amazon (2023).

This approach can be boiled down to four key steps, as illustrated in Figure 14.2. Each of these steps is intuitive. Before we can train Digit to conduct warehouse operations and last-mile deliveries, we must first capture data. You may ask: what kind of data? We need data that captures the diverse environments and scenarios Digit will encounter, both inside and outside the warehouse, along with the corresponding pose, posture, and footstep patterns it must adopt to navigate these scenarios effectively.

So, what is the fastest way to capture this data? To accelerate the data-capture and humanoid-training process, Amazon has developed specialised smart glasses that are provided to its existing team of delivery drivers.[37] These glasses serve two key purposes. First, they provide routing guidance directly on the display located on these glasses, freeing drivers' hands and eyes from mobile phones. The first-order effect of this approach is that it immediately reduces the cognitive and operational burden on the delivery team by simplifying the navigation process.

The second-order effect is that these same glasses enable Amazon to capture vital neighborhood-specific and scenario-specific data, along with the corresponding footstep patterns, postures, and movements that drivers naturally adopt during deliveries. These two data streams can then be used to help Digit understand the types of environments it will need to navigate, as well as the locomotion and navigation strategies it must adopt to successfully complete the delivery process. In simple words, this approach allows Amazon to collect organic data at scale and in a time-efficient manner. These datasets are then used to train the next generation of AI models that Amazon is currently building to enable last-mile deliveries with humanoids.

Amazon has already started testing the Digit humanoid in warehouse operations. Digit can move, grasp, and handle objects of all shapes and sizes across the different areas within these warehouses, overcoming all constraints related to locomotion and spatial manipulation. Its size and shape are well suited for buildings designed for us humans, and Amazon believes there is a significant opportunity to scale a humanoid solution such as Digit, which can work collaboratively with employees. The initial use of this technology will be to help employees with tote recycling. A tote refers to the empty baskets used in warehouses to stack items on shelves and move them along conveyor belts. In the warehouse industry, tote recycling is a highly repetitive process of picking up and moving empty totes from the conveyor belts. Amazon plans to automate this process using Digit.

Beyond Digit, Amazon plans to test other humanoids, including one developed by Unitree and potentially one of the lowest-cost humanoids ever made available to developers, priced at $16,000.[36] Through these trials, Amazon will build its own dataset that will be used to develop a centralized AI model to train its next generation of humanoids.

This concludes the chapter on Amazon's approach to building human-robot interaction systems and humanoids at scale. Amazon and Tesla are not the only industry giants leading the robotics frontier. While warehouses and automotive manufacturers are no doubt first in line when it comes to adopting robots at this scale, the logistics and shipment industry shares the same amount of love for robots as these two industry verticals. The story of Boston Dynamics and its robots is proof of this adoption trend.

Chapter 15: Decoding Boston Dynamics' Approach to Building Spot

Boston Dynamics, a name so familiar that most robotics professionals picture the Spot robot dog or the Atlas Humanoid performing stunning moves in the YouTube world when they read it these days. Few are aware that the history of this project dates back all the way to the 1980s. At the time, Marc Raibert, the founder, was leading a research team on legged robots at CMU and MIT. This research work later culminated in the launch of probably one of the first few robotics companies in the world, i.e., Boston Dynamics, in 1992 with funding from DARPA. Since then, over these last 35+ years, after undergoing a series of acquisitions by a surprisingly diverse array of companies like Alphabet (Google), SoftBank, and now Hyundai, Boston Dynamics has launched some of the most frontier robotics models in the world.[38-39]

From the Atlas Humanoid to the Spot quadruped and the Stretch robot arm, the company is seeing rapid adoption of its technology in the real-world by leading industry-players across diverse sectors such as logistics, manufacturing and energy infrastructure. This list of adopters today includes but is not limited to, Gap, Hyundai, Maersk, National Grid, Michelin, GlobalFoundries, and

Purina. Add to that 1,000 robot orders recently placed by the Dutch logistics giant - DHL.[40-41]

This galactic scale of adoption naturally leads us to this question: what is the core problem that Boston Dynamics is trying to solve with robots. Any guesses?

Mission: Core Problem

Boston Dynamics is on a mission to build robots that can navigate our physical world and take on tasks that are dangerous, repetitive, or physically challenging for us humans. To truly understand this mission and the core problem it's trying to solve, take a pause here and try to visualize the daily lives of the professionals working in oil refineries, mines, automobile factories, and warehouses. These environments place extreme levels of physical and psychological stress on us humans—demands our biological system was never designed to sustain in the first place. Some of you might ask: what are these stress vectors and why are they so significant?

Impact of High-Risk Environments on the Human Body

For starters, here are a few statistics that might help to understand the severity of this challenge, i.e., these stress vectors. Mines, or to be more specific, coal mines exhibit extremely high levels of carbon monoxide gas, i.e., CO. It's one of the most dangerous toxic gases found in these environments. To give you a perspective, prolonged exposure to a CO level of 400 ppm and higher for three to five hours will result in convulsions, coma, brain damage, and even death. Now to put this risk profile of mines for human workers into perspective, it helps to understand that there are about 21,000+ active mines around the world today, each employing 1000s of workers.[42,43] Each worker is at a risk if inhaling these gases for hours a week. Can you imagine the sheer amount of stress this puts on the brain and the respiratory system of these workers?

Thus, the 86 billion neurons located in our brain are not designed to handle the sheer amount of physical and mental stress resulting from working in these highly toxic environments. The same principle applies to our respiratory system. The 300 million alveoli—i.e. an array of tiny sacs located in our lungs that enable us to breathe 24,000 times a day—are solely designed to enable the gaseous exchange of oxygen and carbon dioxide.[44] Forcing this system to handle toxic gases and pollutants for prolonged durations puts our respiratory system under extreme levels of stress. In summary, prolonged exposure to toxic emissions in oil refineries and mines, or the intense physical strain resulting from lifting heavy industrial parts for hours in factories, can severely impact our brain, joints, and muscles.

These are often the risks that are frequently overlooked in the industrial environments—risks that make robots a logical and necessary solution, not just in the long term but in the near future. This is precisely the mission that drives Boston Dynamics and many other robotics companies: to build autonomous machines that can operate safely and effectively in environments that are hazardous or unsustainable for humans.

This mission began over 30 years ago in the research labs of CMU and MIT and laid the foundation for Spot, their agile quadruped robot. The company then expanded into spatial manipulators with Stretch, a mobile robot arm designed exclusively to automate warehouses and logistics operations, and is now building the next-generation humanoid robot, i.e. Atlas. This pace of robot-led innovation naturally leads to another question: so, who is using these robots, where, and how? To answer that question, we will start with Spot, their quadruped robot.

Many of you have probably seen this yellow quadruped robot, illustrated in Figure 15.1, performing stunning moves on social media platforms. It is not a recent entrant in the legged-robot space. Its history dates back some 30 years, when DARPA funded this legendary robot project conceptualised by research teams at CMU and MIT. This initial research on legged robots led to the development of massive, hydraulically powered robot dogs—namely BigDog in 2004—and over the decades, the project evolved into a compact, battery-powered, electrically actuated platform: that's Spot.[45]

Today, Spot weighs around 33.8 kg (74.5 lbs), can carry a 14 kg (30 lbs) payload, runs for about 90 minutes on a full charge, and achieves speeds of up to 1.6 m/s (3.5 mph).[46] It is not the only quadruped robot on the market—companies like Unitree Robotics, with their Go1 series, have developed similar platforms. Spot, however, remains the gold standard for rugged legged robots designed specifically for industrial and outdoor environments. Why is that the case? Why not use mobile robots that dominate warehouse floors? And since we are talking about legged robots, why not use humanoids?

Spot — The Quadruped Explorer

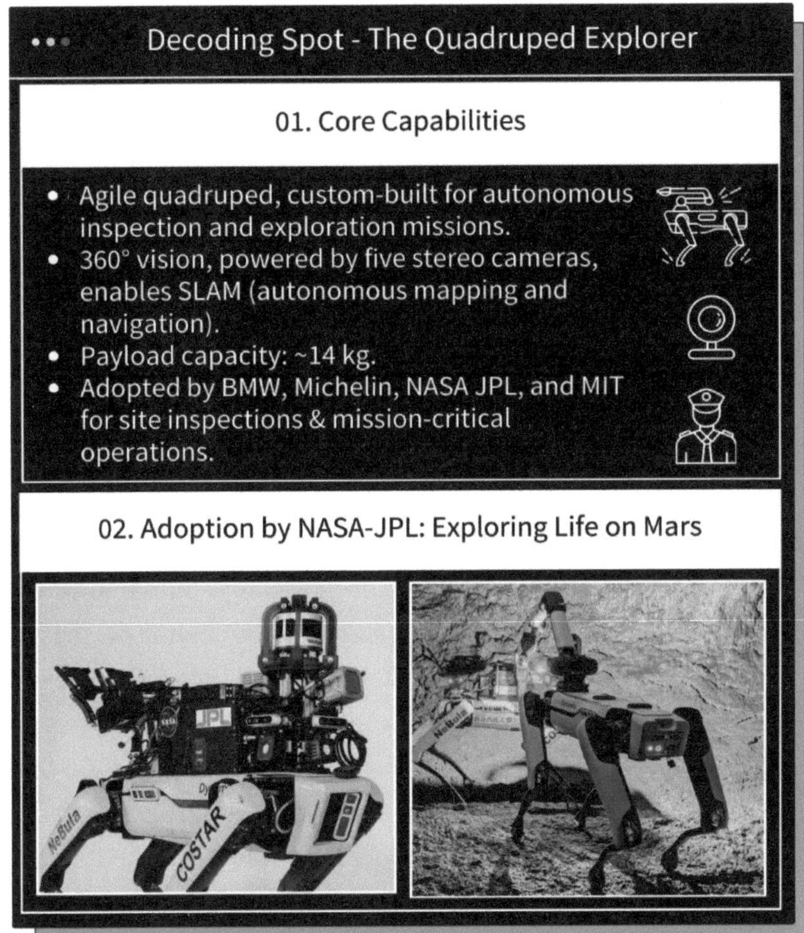

Figure 15.1 — Decoding Spot — The Quadruped Explorer

Illustration by the author. Spot robot image from the Boston Dynamics YouTube channel, "Spot Watches Its Step" (2023), and from "Robotics at JPL — NeBula-SPOT," Jet Propulsion Laboratory (2023).

Why Quadrupeds Outperform Mobile and Humanoid Robots on Diverse Terrain Profiles

The answer here really boils down to the robot's ability to adapt to diverse terrain profiles. While mobile robots offer extremely high speed and payload-carrying capacity, try programming one to navigate slopes, staircases, and escalators. Try programming one to navigate hills, valleys, mountains, and deserts. It is a nightmare, to say the least. Make no mistake—mobile robots perform very well if you are designing one for a given terrain profile. But if you need a robot that can perform across diverse terrain profiles, that may not be the best solution. That naturally leads to the next question: why not deploy humanoids instead of robot dogs?

Again, humanoids are good at navigating diverse terrain profiles, but when you transition from robot dogs towards humanoids, you realise you now need to factor in a significantly high-risk variable in your robot equation—balance. Retaining balance for a 5 ft 4 in, 120 lb. robot in deserts, on slopes, and across mountains is a great research project for laboratories but certainly not worth the time and resources for a robotics startup. Thus, quadrupeds offer a balanced solution that falls between mobile robots and humanoids. These quadrupeds thrive in ecosystems that are harsh, high-risk, or inaccessible to humans or wheeled robots—deserts, forests, chemical plants, nuclear power stations, offshore oil rigs, and underground mines. Hence, in such high-risk zones, Spot's key value proposition really boils down to terrain adaptability and remote inspection.

In fact, most of Spot's commercial deployments today revolve around industrial inspection projects. For example, British

Petroleum uses Spot on oil rigs to conduct remote inspection and monitor gas leaks—especially methane leaks, which are both dangerous and common on oil rigs. NASA's Jet Propulsion Laboratory has tested Spot in underground caves here on Earth, while the Los Angeles City and New York Fire Departments have adopted it to augment existing public safety and emergency-response systems.[41] With each deployment, organisations gain the ability to collect high-fidelity data from these high-risk environments, allowing human inspectors managing these sites to address corresponding safety and security risks in a timely and resource-efficient manner.

This naturally leads to our next question: how does Spot work?

Decoding Spot's System Architecture

Spot's operations can be distilled into five key steps, as illustrated in Figure 15.2. They cover perception, motion planning, balance retention, execution, and optimization.

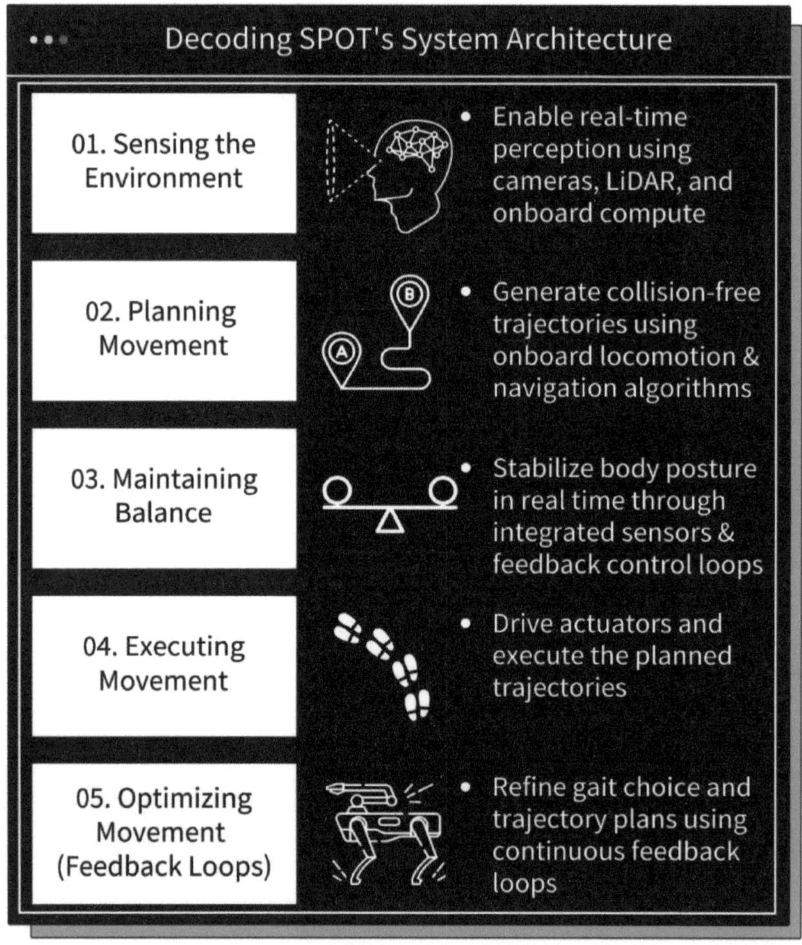

Figure 15.2 — Decoding SPOT's System Architecture

Illustration by the author.

Today, it can operate in two modes as illustrated in Figure 15.3.

Spot's Mode of Operation: Manual vs. Autonomous

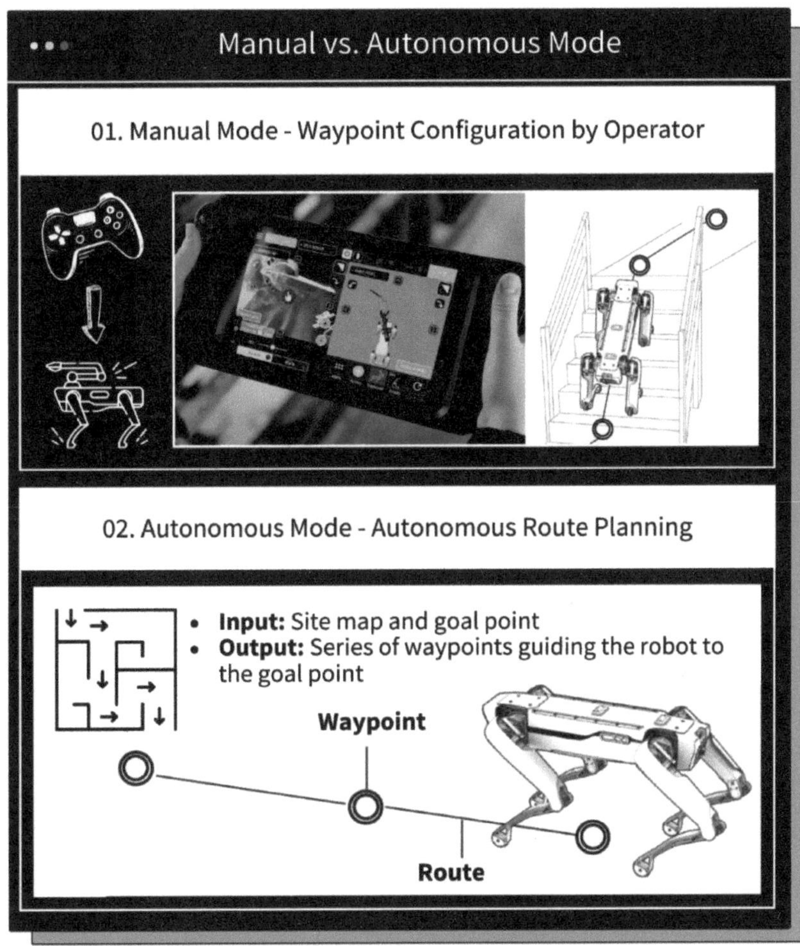

Figure 15.3 — Manual vs. Autonomous Mode

Illustration by the author. Spot robot image from the Boston Dynamics YouTube channel, "Spot Watches Its Step" (2023).

In manual mode, an operator sends commands to Spot using either a joystick or a web-based controller. These commands are then converted by the onboard computer into control signals for the actuators, which in turn drive the robot.

In autonomous mode, assuming Spot has already mapped its environment or, if operating outdoors, has access to GPS coordinates, it is provided with a goal or destination point. From there, it develops an end-to-end route from its current position to that point and completes its mission accordingly.

Now that we understand the two modes of operation for Spot, let us explore the five key steps that are critical for Spot to complete its mission. We will cover each of these steps one by one.

Step I – Sensing the Environment

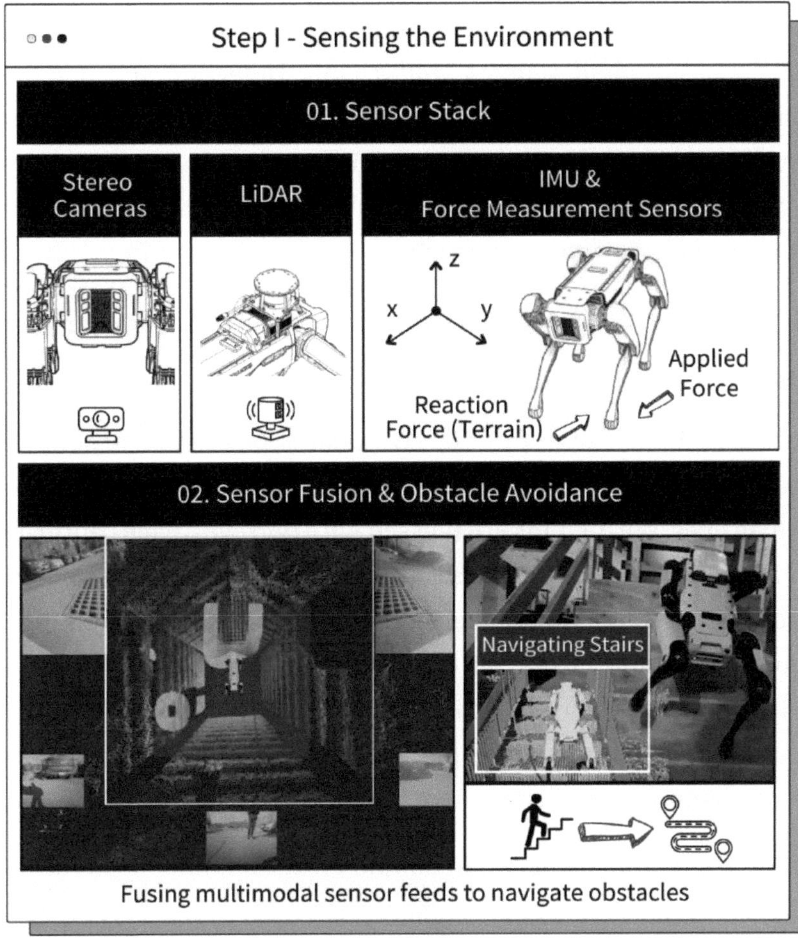

Figure 15.4 — Step I – Sensing the Environment

Illustration by the author. Spot robot image from the Boston Dynamics YouTube channel, "Spot Watches Its Step" (2023).

The first step here is to sense the environment. In this step, as illustrated in Figure 15.4, the Spot robot uses its onboard sensors to develop a coherent view of its surroundings. To enable this, Spot is equipped with a set of perception sensors, including but not limited to stereo cameras, inertial measurement units (IMUs), force measurement sensors, and a Light Detection and Ranging sensor (LiDAR). It is this array of sensors that allows Spot to develop a coherent view of its surroundings.[47] That naturally leads to the next question: how does it develop this coherent view? We will start with the stereo camera array.

How Does Spot Develop a Coherent View of its Surroundings?

Spot is equipped with five stereo cameras positioned around its body frame. For starters, a stereo camera functions much like human eyes. We humans are blessed with two eyes, each capturing an image of our surroundings. Our brain then superimposes these two images to create a coherent view of the environment. This superimposition allows the brain to perceive depth, helping it detect how far each object is from our eyes within that unified image. This phenomenon is known as stereo vision, and the five stereo cameras installed on Spot function in exactly the same way.

Those five stereo cameras, in simple terms, amount to about ten RGB cameras. Each stereo camera internally contains two RGB cameras, each capturing a single image of the surroundings. These two images are then superimposed, allowing each stereo camera to provide Spot with both visual and depth information from the scene. While cameras do a good job of providing Spot with a clear view of its surroundings, they are light-dependent sensors. This means that at night or under poor lighting conditions, they cannot provide the level of accuracy and precision needed for Spot to sense its environment. That is where the next set of sensors, i.e., IMUs and LiDAR, comes into play.

IMUs enable Spot to measure its orientation and acceleration in 3D space. This measurement is important for Spot to track its exact position as it navigates the physical world. Beyond the IMU, Spot is also blessed with the LiDAR sensor. Unlike cameras that are dependent on light conditions to provide Spot with a unified view of its surroundings and distance to each obstacle, LiDARs do not share the same downside. LiDARs not only provide a much better visual range than cameras to Spot but also eliminate visual blind spots during night-time and under poor light conditions. This combination of cameras and LiDARs offers Spot a mission-critical vision system, as illustrated in Figure 15.4. Needless to say, since the team is using an array of vision sensors here, the data from each of these sensors is combined using Kalman filters and deep-learning-based fusion models to ensure Spot gets a unified view of its surroundings.

While this vision system enables Spot to develop a strong sense of terrain and to plan a trajectory for navigation, without a feedback system in place, it becomes an open-loop system. In other words, unless Spot falls, there is no way for it to detect how effective its navigation plan is. So, how can we collect feedback from the environment at each step that Spot takes from its source to its destination point?

That is where the force measurement sensors come into play. The team building Spot has deployed these sensors on its legs to ensure they are able to accurately measure the impact of the ground on each of its legs and then extrapolate the cumulative effect of this impact on Spot's ability to retain its balance.[48-49] This feedback can then be used to optimize each step Spot takes and deliver a robust performance. Thus, this vision system enables Spot to develop a coherent view of its surroundings. Vision is just step one in this operational equation of Spot.

Next, it needs to develop a detailed navigation plan to move safely from its source point to its destination or goal point. How does it do that? That is the function of the navigation planner.

The process of developing a detailed plan to navigate from the source to the destination point involves three key sub-steps: path planning, footstep planning, and gait planning, as illustrated in Figure 15.5. Let us cover each of these steps one by one.

Step II – Planning Movement

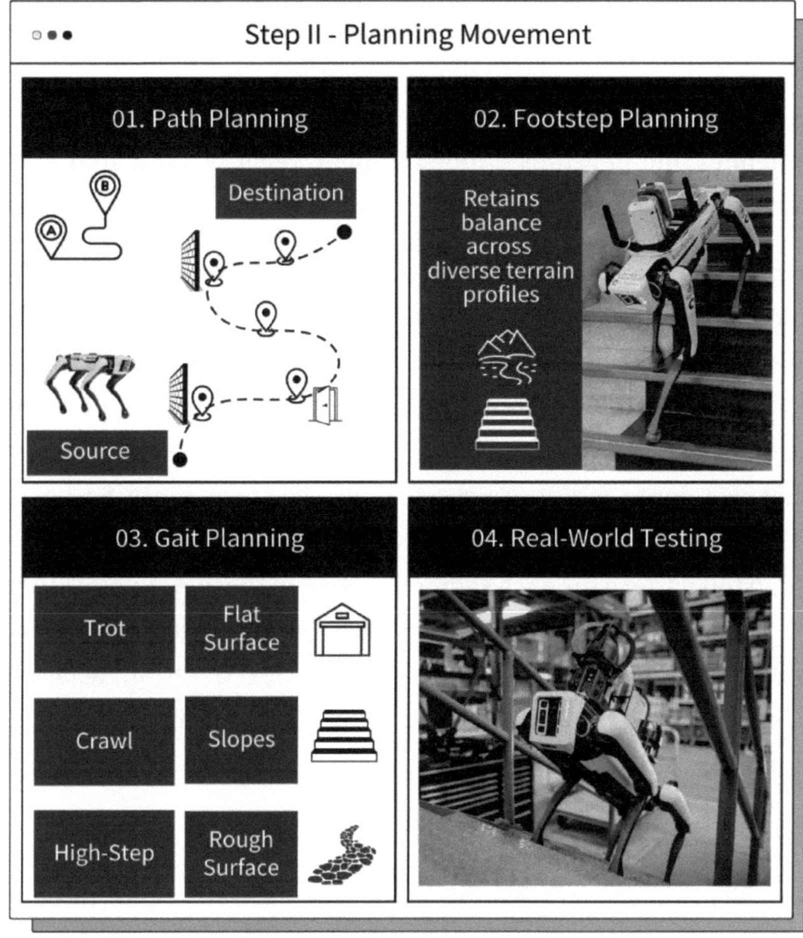

Figure 15.5 — Step II – Planning Movement

Illustration by the author. Spot robot image from the Boston Dynamics Spot product page (2025)

First, Spot plans an end-to-end route, or a global plan, from the source point to the destination point. To do so, if it is in an indoor space, it uses the standardised global planner provided by ROS, i.e. the Robot Operating System. On the other hand, if it is in an outdoor space, it relies on a GPS-based route planner. Once it develops this global plan, which in simple terms boils down to a series of waypoints generated by the algorithms, it uses those as guidelines and develops a footstep plan. What is a footstep plan, and why do we need one?

Remember, legged robots or more specifically quadrupeds, function in a totally different manner from mobile robots. In the case of mobile robots, the decision to follow the waypoints pretty much boils down to turning the wheels forward or in reverse. In the case of legged robots, there are a series of footstep patterns, or what we refer to as gait choices, that we need to factor in. So, based on the terrain profile, slope, grip, and other factors, Spot develops a footstep pattern for each segment between these waypoints with the aim of retaining balance at all costs while marching towards the destination. That naturally leads to the next question: what are these gait patterns? Also, how many gait patterns does Spot support?

Gait Patterns Exhibited by Spot

In reality, quadrupeds exhibit anywhere between three to eight gait patterns. For simplicity, we will stick with the top three, i.e. Trot, Crawl, and High-Step. In the Trot mode, two diagonally opposite legs are operational at all times. This movement resembles that of a dog or cat walking on a planar surface. This gait choice is best suited when Spot needs to navigate across flat or mildly rough surfaces. While Trot mode is good for flat surfaces, if it has to navigate a slope or uneven surfaces, it needs to adopt a different gait choice. That is where the Crawl gait comes in handy. It offers a relatively slow yet highly stable movement for surfaces that demand higher grip, such as stairs, slopes, and highly uneven terrain profiles.

While Trot is good for flat surfaces and Crawl for extremely uneven terrain, what about debris, pipes, and surface profiles that are neither 100% flat nor extremely uneven? That is where the High-Step gait comes in. Thus, Spot develops a footstep plan as it navigates from its source towards its goal point.[48-50]

Now, legged movement is inherently unstable. No matter the gait choice, we need a mechanism in place to ensure Spot is able to retain its balance at all times. How do we do that?

That is where modern reinforcement learning algorithms play a critical role.

Step III – Retaining Balance (MPC vs. Reinforcement Learning)

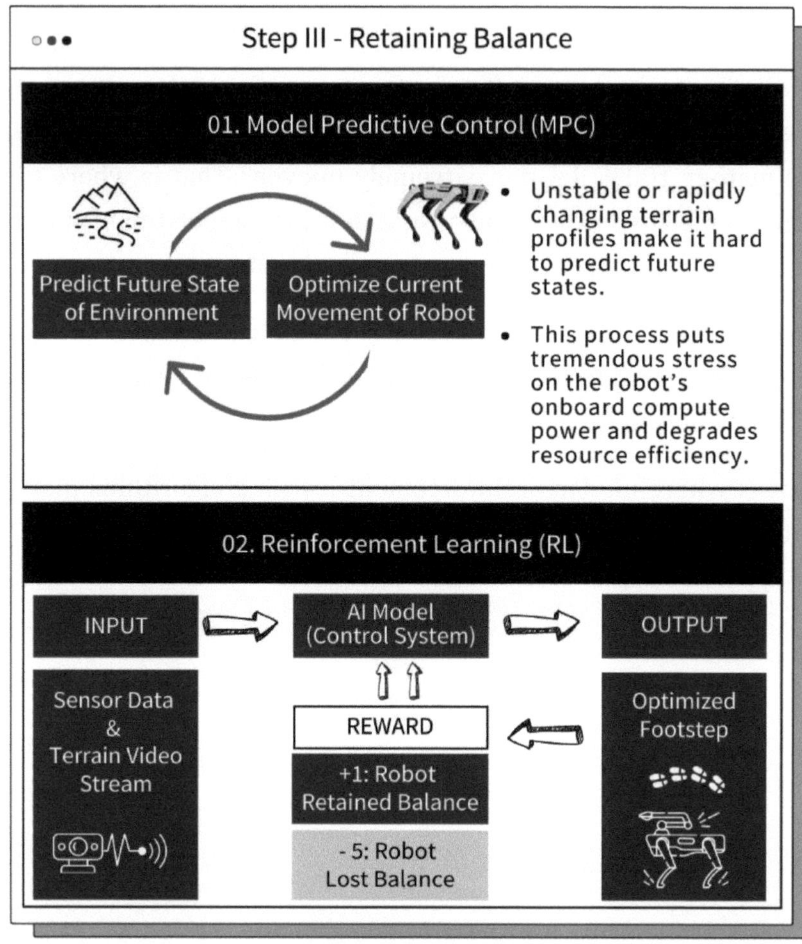

Figure 15.6 — Step III – Retaining Balance

Illustration by the author.

Retaining balance is the birthright of all legged robots, and Spot is no exception. The team designing Spot incorporated two approaches to handle this challenge, namely Model Predictive Control (MPC) and Reinforcement Learning (RL), as illustrated in Figure 15.6.[48-49]

Model Predictive Control (MPC) helps Spot predict its next physical state and the terrain ahead based on the real-time data collected by the sensors. For example, if at time t = 1 second the terrain is flat, the system might assume, based on sensor readings, that by time t = 3 seconds, the terrain will remain largely unchanged. This means the MPC controller would likely select the same gait or leg movement pattern it used earlier when the ground was flat. But here is the catch: what if, at t = 3 seconds, the terrain has completely changed, say the robot has stepped from grassland into a shallow river? You see, while MPC offers important benefits, it also has limitations. MPC, by design, assumes a somewhat linear and predictable world. This assumption holds well inside controlled laboratory environments but often fails when the robot is introduced to the real world.

That is where Reinforcement Learning (RL) becomes valuable.

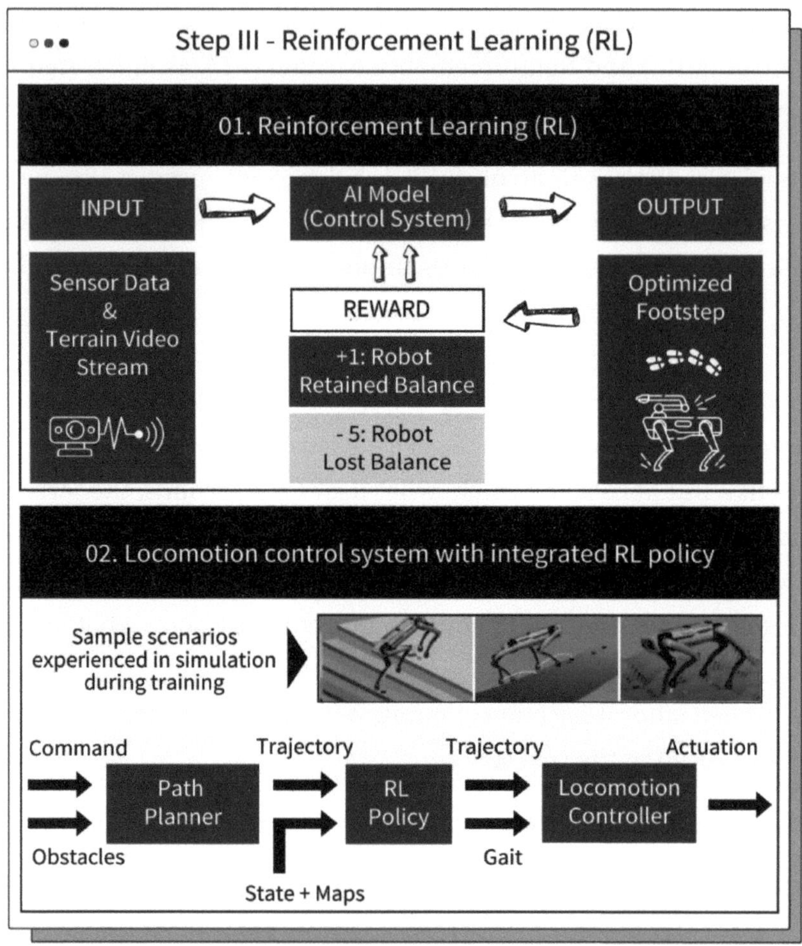

Figure 15.7 — Step III – Reinforcement Learning (RL)

Illustration by the author. Spot robot image from the Boston Dynamics blog, "Starting on the Right Foot with Reinforcement Learning" (2023).

RL allows Spot to learn stable navigation patterns through a structured trial-and-error process, as illustrated in Figure 15.7. In a simulation environment, Spot is trained to maximise a reward function that promotes balance retention, energy efficiency, and progress towards a goal, while penalising slips, falls, or excessive energy consumption. Over millions of iterations, this process yields a policy—essentially a learned reflex—that allows Spot to respond fluidly to changes in terrain without requiring an explicit model of the environment. In practice, this RL-derived policy often demonstrates superior performance to traditional MPC methods, especially when Spot is tested in unstructured, real-world environments.[48-49]

What does Spot do next once it learns this policy?

Step IV & V – Executing and Optimizing Movement (Motion Control and Feedback Loops)

Once the navigation system has determined the best path to reach the destination, the motion control system takes over. This subsystem translates the high-level trajectory developed by the navigation planner into precise motor commands that regulate torque, velocity, and joint position across all twelve leg motors. Executing the plan is one part of Spot's equation; optimising it is the next. But why?

While the high-level navigation plan developed by Spot's planning system allows it to march towards its destination, it is equally important for Spot to sense feedback from the terrain with each step it takes and make fine adjustments to the next one. This ability ensures that Spot responds almost immediately to even minor changes in the ground surface, preventing unexpected slips or falls. How does it do that? This is where the optimization control loop comes into play.

This control loop operates at extremely high frequencies, often hundreds of cycles per second, enabling Spot to react almost instantaneously to dynamic obstacles that may cross its path and abnormal terrain shifts that it might encounter with each step. Each motor receives continuous feedback from onboard sensors that monitor a wide range of motion vectors associated with each leg of Spot, such as position, force, and acceleration, enabling micro-adjustments to retain balance and ensure stable operations at all times. This closed-loop control ensures that Spot's movements remain fluid, stable, and energy-efficient, even across uneven or unpredictable terrain.

That summarizes the operations of Spot. Spot is not the only robot developed by Boston Dynamics that is experiencing galactic adoption across the globe. Stretch, its mobile robotic arm, is already helping logistics giants such as DHL transport and deliver thousands of packages each day. How is it doing that? We will explore its story in the next chapter.

Chapter 16: Decoding Boston Dynamics' Approach to Building Stretch

Stretch — The Industrial Robot Arm

In the previous chapter, we learned how a project that started in the research labs of CMU and MIT culminated in the launch of three industrial robots namely Spot, Stretch, and Atlas that are now being adopted around the world. From Gap to Michelin, British Petroleum to National Grid, and Hyundai, these robots are gaining wide-spread adoption across industries. We covered Spot in the previous chapter and learned how it aids in industrial monitoring, particularly in environments that are high-risk for humans, such as mines, tunnels, factories, and oil refineries. While Spot excels at conducting remote patrols in these spaces, it's not designed to lift heavy payloads. Why does that matter?

The Need for Developing the Stretch Robotic Arm

If you recall, the mission of Boston Dynamics is to build robots that can assist humans with tasks that are dangerous and/or physically demanding. Patrolling a mine or tunnel is indeed dangerous and physically demanding, but it doesn't necessarily involve lifting heavy payloads. There are already specialized machines designed for those environments, so Spot doesn't need to handle heavy payloads.

That's not the case in logistics, warehouses, and factories. The majority of heavy work in these industries boils down to a company's ability to move goods and packages at scale and at maximum speed. While trucks and conveyor belts provide some assistance in these spaces—both outside and inside these facilities—human workers are still dedicated to a series of tasks that involve handling goods and packages of all shapes and sizes. This is especially dangerous and physically demanding work.

That is precisely the problem Boston Dynamics wanted to address, following its success with the first use case, i.e., remote patrol, which we covered in the previous chapter. This core challenge led to the development of its mobile robot arm called Stretch. Essentially, Stretch is an extremely adaptive industrial robotic arm mounted on an autonomous mobile base and designed exclusively for industrial environments.

Some of you may ask: there are hundreds of industrial robotic arms available today, so why build another one? What makes Stretch novel?

Why Traditional Robotic Arms Struggle in Modern Warehouses

To better understand what makes Stretch novel, it helps to take a step back and look at the challenges of deploying an industrial robotic arm in a warehouse or logistics facility. A vast majority of robotic arms available today in the industry are static, meaning they need to be permanently bolted to the floor of these factories and warehouses. That's a big problem. Why?

These environments, at the end of the day, are responsible for producing or processing large volumes of products and packages. This means that, based on demand and supply requirements, the internal footprint of these facilities changes almost every month, if not every week. Now, if you bolt an industrial robotic arm to the floor, it becomes extremely difficult for most logistics and warehouse operators to reorganise their layouts to better serve these dynamic demands. This was indeed a major problem for customers like DHL and Maersk.

That's not the only challenge. These customers also needed a robotic arm that could adapt efficiently to a wide range of goods and packages that come in all shapes and sizes. So how did Stretch address this challenge?

How Stretch Solves the Mobility and Adaptability Challenge

The team building Stretch understood these challenges from the very beginning and therefore deployed the robotic arm on a custom-designed autonomous mobile base. This mobile base gives operators the ability to move Stretch with ease within factories and warehouses and, in many cases, directly to shipping containers so it can operate effectively within confined spaces. That solves the mobility challenge.

To address the second challenge associated with handling goods and packages of different shapes and sizes, the team equipped Stretch with high-performance precision grippers. Each of these arms is equipped with an array of advanced grippers capable of lifting a wide-range of payloads.

This unique engineering approach gives Stretch extraordinary spatial agility. You can deploy it in one part of the warehouse today and move it to another part tomorrow, as needs change, within minutes. Similarly, you can use it to handle packages of all shapes and sizes—think books to auto parts, appliances to culinary equipment. It can handle packages across a wide range of industry verticals. This is a completely different approach from that adopted by developers of 99% of industrial robotic arms we see in factories today, those that remain bolted to the floor. Stretch provides this rare combination of mobility and dexterity.

Beyond this mobility, Stretch is also one of the strongest mobile robotic arms in the industry, capable of handling a staggering 800 cargo cases per hour.[51] How does it do that? That's where the engineering approach employed to build Stretch's comes into play.

Core Capabilities: Engineering for Tight Spaces and Heavy Payloads

If you carefully observe the Stretch mobile robot arm illustrated in Figure 16.1, you'll notice that it is purpose-built for space-constrained environments such as trailers, loading docks, and container yards. This means it must adhere to the strict demands imposed by these environments. What are these demands?

First in line is the compact footprint. You cannot put a bulldozer-sized robot arm inside a trailer or near a loading dock. Doing so would negatively impact both the performance of the robotic arm and the customer's ability to move it in and out of trailers and docks. Hence, you need a robotic arm that offers a compact footprint and extremely high mobility.

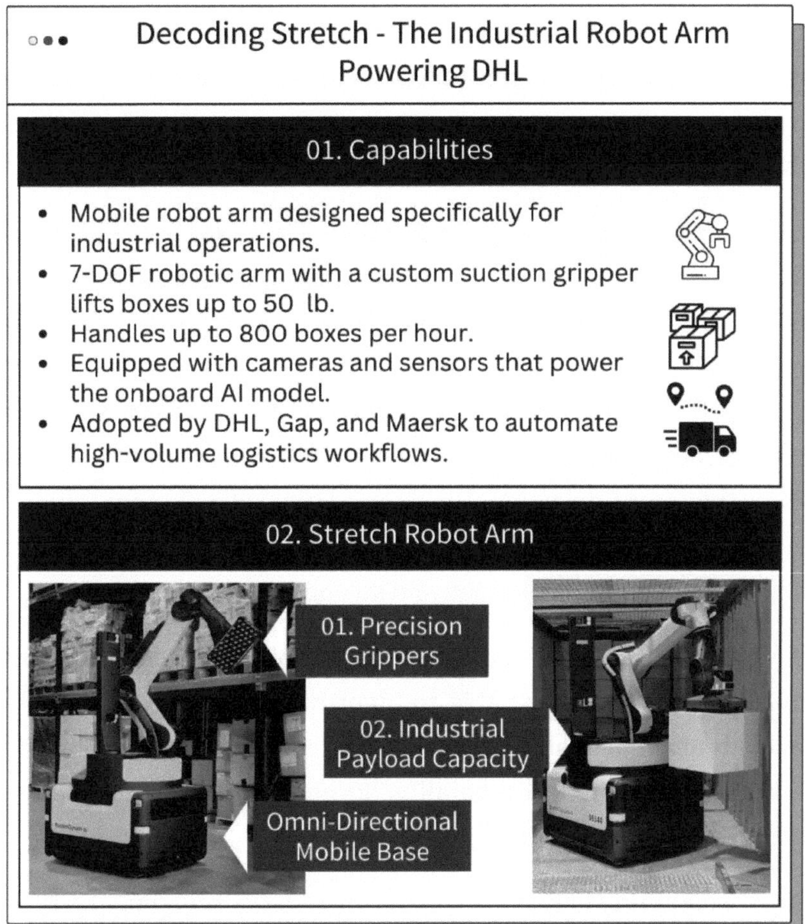

Figure 16.1 — Decoding Stretch — The Industrial Robot Arm Powering DHL

Illustration by the author. Stretch robot arm image from the Boston Dynamics product page, "Stretch" (2025).

To meet these requirements, as illustrated in Figure 16.1, Stretch is built on an omni-directional base powered by holonomic motor drives. Unlike the steering systems of most modern cars, which prevent you from turning them in one place, the omni-directional base enables Stretch's operators to rotate the entire arm assembly on the spot without enforcing wide turn radii. This is a major advantage for operators working in such environments. Why?

Stretch has to operate in extremely tight spaces and work alongside conveyor belts. Try to visualize operating a robot arm inside a 53-ft trailer where space is tight and turning radii are limited. This is where the holonomic motor drives give the ability to Stretch to move in all directions while retaining an ultra-minimal footprint. This design choice is not cosmetic—it's critical. While this omni-directional base powered by holonomic motor drives solves the footprint challenge, it doesn't address the next challenge, i.e. spatial coverage and strength.

An industrial robot arm designed to operate in these environments should be able to identify, lift, and transfer heavy payloads at a reasonable pace for extended periods while retaining mobility, precision and reliability. So how did the team building Stretch address this challenge?

They custom-designed a seven-degree-of-freedom (7-DOF) robotic arm and mounted it on that omni-directional mobile base, as illustrated in Figure 16.2.

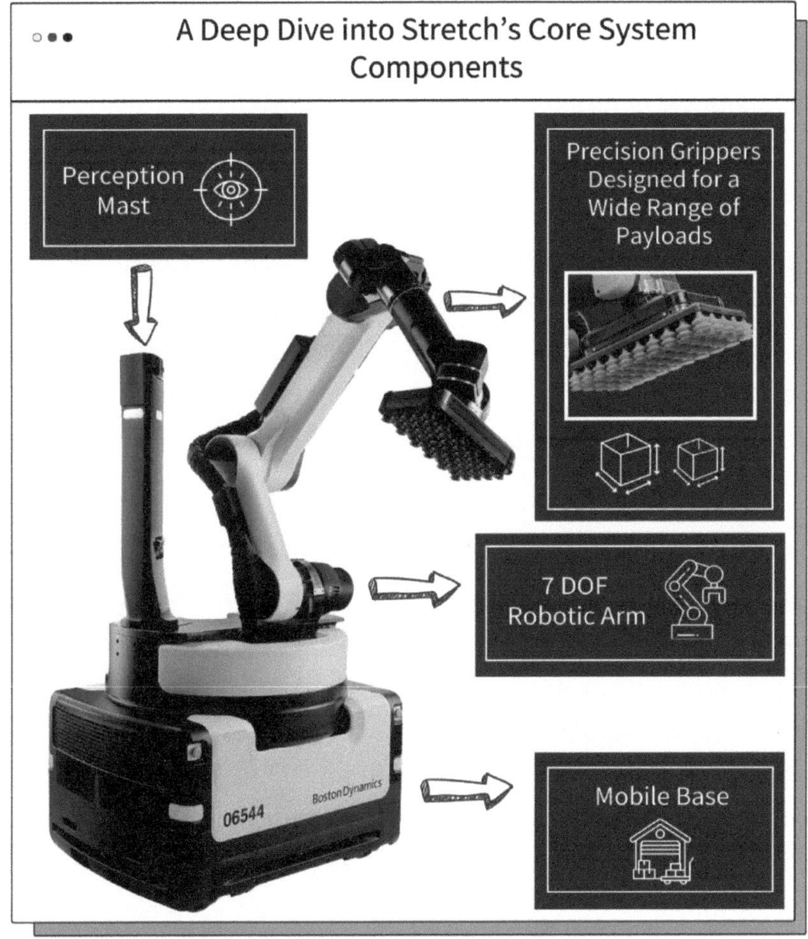

Figure 16.2— A Deep Dive into Stretch's Core System Components

Illustration by the author. Stretch robot arm image from the Boston Dynamics product page, "Stretch" (2025).

This custom-designed 7-DOF arm gives Stretch extended coverage, strength, and speed. The arm extends to approximately 10.5 feet (3.2 metres), allowing it to access both the ceiling and floor of standard trailers with ease. At the end of this arm is a high-precision vacuum gripper capable of handling boxes weighing up to 50 pounds. In optimal configurations, Stretch can transfer up to 800 boxes per hour from trailers to conveyor belts or vice versa, operating continuously for hours without interruption.[51-52]

This ability to operate for long hours naturally leads to the next defining characteristic of Stretch: endurance. Stretch is designed for industrial-grade endurance. A single charge enables it to operate multiple shifts, typically 12 to 16 hours depending on the setup, making it suitable for round-the-clock warehouse operations.

From a perception and autonomy standpoint, Stretch is equally sophisticated. It features a vision mast, an array of cameras and sensors paired with AI models that analyze the surrounding scene in real time. This system allows Stretch to detect packages under varied lighting conditions, estimate their poses accurately, and transfer an array of them at a lightning speed, thereby ensuring maximum throughout.[52] It can even recover autonomously from failures. For instance, if it drops a package, it recognises the error and, in most cases, recovers without manual intervention. The underlying AI model has been trained to ensure it learns with each failure and improves along the way.

Beyond endurance and perception, the next question some of you might ask relates to fleet management. Stretch is rarely deployed as a single unit. A customer typically purchases at least a dozen, if not hundreds, of these robots. So how do they manage this fleet of robot arms?

Coordinating multiple Stretch robots at scale requires more than good mechanics and a vision system. Stretch comes with a fleet operations system that enables logistics operators to manage and schedule several units simultaneously, ensuring that each robot fits seamlessly into their existing workflows. When you work with Stretch, you realize that it's strategically handcrafted to operate with maximum efficiency under warehouse constraints—tight spaces, around-the-clock operation, heavy payloads, and the need for seamless integration with a wide range of conveyor systems.

This is exactly why major logistics and industrial giants such as DHL, Gap Inc., Otto Group, and H&M are adopting Stretch to scale their dock operations. How are they doing that? We will cover that in the next chapter.

Chapter 17: How DHL, Gap, & Maersk Are Adopting the Stretch Robot Arm

Adoption at Scale: Who's Using Stretch and Why It Matters

Stretch's primary use case today is anchored in the warehouse and logistics industry. If you look closely at the adoption data, thousands of these units are transferring packages to and from trailers and containers at a consistent pace for hours across the globe, addressing a trifecta of industry pain points: seasonal staffing shortages, injury risks arising from repetitive lifting of heavy payloads, and unpredictable container unloading speeds. Beyond addressing these pain points, it is explicitly designed for warehouse environments—it can be set up alongside a wide range of conveyor belts, requires minimal site modifications, and can be operated by a warehouse professional after just a couple of days of training. With the ability to handle a staggering 800 cases per hour, each weighing up to 50 lbs., and capable of operating across multiple shifts on a single charge, Stretch monastically fits into the existing workflow of the logistics and warehouse industry. And the industry giants have noticed this product-market fit.

DHL's Early Bet on Stretch

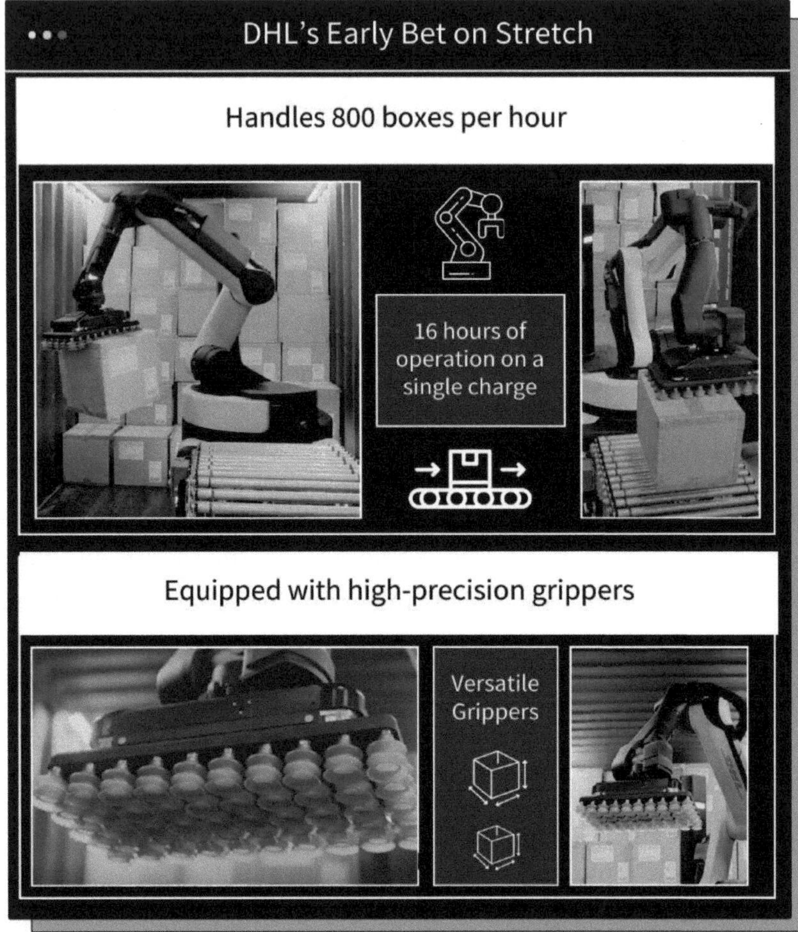

Figure 17.1— DHL's Early Bet on Stretch

Illustration by the author. Stretch robot arm image from the Boston Dynamics case study, "Stretch at DHL" (2025).

The clearest signal of Stretch's traction comes from DHL, one of the world's largest logistics providers. It transports roughly 44 million shipments each year around the globe, and the company became

Stretch's first commercial customer in 2022, following a USD 15 million multi-year deal. By January 2023, DHL had the robot running in live production environments, autonomously unloading trailers in its U.S. distribution centers. The company reported unloading speeds of up to 700 cases per hour, which is at or above the throughput of experienced human teams operating under ideal conditions in these warehouse environments.[53-54]

Continuing this momentum, in May 2025, DHL doubled down on its adoption of Stretch, signing a Memorandum of Understanding with Boston Dynamics to deploy more than 1,000 Stretch robots worldwide.[40] This DHL agreement represents one of the largest commitments to mobile robot arms ever announced. That naturally raises the question: what made the team at DHL adopt these robots at such a scale? When asked about this adoption curve, the DHL team pointed to two decisive factors: time-efficient robot-setup procedures and minimal operator-training requirements. In many cases, Stretch can go from delivery to full operation in just "a couple of days," demonstrating that it can be integrated seamlessly into existing docks without the need for complex retrofits.[55] DHL may have been first to adopt mobile robot arms at this galactic scale, but it is certainly not alone.

Gap, Otto, and Maersk Join the Stretch Revolution

Gap Inc. also deployed Stretch robots in 2023, automating trailer-unloading operations across parts of its distribution network. Meanwhile, major European & North American logistics and retail players—including Otto Group, Maersk, H&M, Arvato, and NFI—have joined the customer roster.[41]

Otto Group offers one of the most transparent looks at real-world performance. At its Haldensleben logistics site in Germany, Otto initially targeted a throughput of 500 packages per hour. In practice, Stretch has exceeded expectations, consistently reaching approximately 550 cases per hour. The company has already expanded its deployment to three robots at that site and has publicly stated its intent to roll out over 100 units across its logistics network in the coming years.[41]

Beyond throughput, Otto notes that the time required to commission new robots at its sites has decreased significantly over time. This also means employee roles are evolving at a similar pace. Employees are being upskilled to manage these fleets of robots and scale operations accordingly.

That naturally leads to this question: how many packages have been handled to date globally by these robots? What does this level of adoption look like on a global scale?

2 Million Boxes and Counting

Around August 2023, Boston Dynamics announced that Stretch had moved more than one million boxes across all customer sites in just a year.[56] Just four months later, in its year-end review, the company reported that this number had doubled to over two million, driven in part by newly launched software-enabled features like multi-pick, which allows Stretch to grab multiple boxes at once and substantially improve effective throughput. In press coverage of DHL's U.S. operations, field tests consistently show Stretch achieving a throughput between 550 and 700 cases per hour, depending on trailer conditions and box type. These are no longer lab benchmarks—they're numbers from active, high-volume logistics sites. Not one but many, and across all industry verticals.

One Robot, Many Verticals

What's particularly notable is the range of industrial environments where Stretch is gaining traction. DHL represents a global third-party logistics leader. Otto Group and H&M bring European retail and fashion distribution verticals to this pack. Maersk reflects maritime logistics. Gap Inc. represents the large-scale U.S. apparel fulfillment industry. So, what makes Stretch such a versatile robot arm for adoption across these verticals?

The robot's mobility, spatial coverage, and adaptability to the diverse real-world footprints of trailers and warehouses across the globe have made it extremely appealing across these varied verticals. This diversity suggests that Stretch is one of the very few mobile robot arms in the industry capable of autonomous case handling and automating dock operations at scale with precision, reliability and ultra-high dexterity.

To say this adoption curve will remain exponential for years to come is a bit early, but the trend lines here are accelerating at an unprecedented pace. Simply put, Stretch is not yet a "hundreds of customers" product, but the scale of individual deals, such as DHL's 1,000-robot memorandum of understanding, indicates serious commercial momentum. Each deployment leads to more training data, better AI models, and improved economics, creating a feedback loop that makes future deployments even more valuable.

Perhaps most importantly, these deployments also demonstrate the potential of a human-robot collaboration system operating at a global scale and addressing real-world challenges. For decades, Boston Dynamics was known for prototypes that made headlines for ultra-human stunts but struggled to ship a product that solved a real problem at scale. Stretch flips that narrative. It's quietly transforming some of the most complex and labor-intensive nodes of our global supply chain. Its adoption story isn’t just about a single robot; it represents a cosmic shift in how the warehousing and logistics industry is approaching one of its last major automation frontiers. If DHL, Otto Group, and others are any indication, that shift is already well underway.

Now that we have studied the approaches adopted by Boston Dynamics, Tesla, and Amazon, leaders in the robotics industry , to bring modern robots to life, in the next few chapters, we will cover the core lessons and the corresponding frameworks that you, as a founder or engineer operating in this space, will need to establish a strong foothold in the industry.

Chapter 18: Career Traps to Avoid in Robotics

If you're a student or an engineer trying to break into robotics, there are a few things worth knowing before you get too far in. Most of what follows in this chapter comes from mistakes I've made myself, along with patterns I've seen in students who complete a BS or MS in engineering and still struggle to find meaningful roles in this field.

I've split this particular sub-section on career traps into two chapters. This chapter focuses on the traps students fall into during the early stages of their career in this field. The next chapter covers the traps that appear once they begin their job search.

My hope is that these pointers help you avoid the detours that slow so many people down, and, worse, often come at a serious cost in both time and capital. This journey into robotics is hard enough; there's no reason to make it harder by getting stuck in these traps that are surprisingly easy to avoid once you know they exist. If you understand these patterns early and learn to spot them, you can steer your ship in the right direction and build a long-term trajectory in this domain that you can be proud of.

So, what are these traps? Let's cover each of them one by one.

01. Don't Fall in Love with Python, ROS, or Robot Projects

While it is understandable that most students start their journey in robotics by coding beginner-grade robots and subsequently mastering ROS and other programming frameworks, the risk that most run into is that they never invest their time and effort in understanding the challenges of building and deploying robots in the real world.

In simple words, they never take time to understand how companies around the world are building and deploying robots at scale, the risks they are running into, and the corresponding solutions that are being developed. You may ask: why is this exercise important? Well, if you are going to work in this field as an engineer or a founder, you, at the end of the day, will be solving problems. Now, to solve any problem in the commercial world, you need a solution that is extremely optimal in terms of time, capital, and resources.

If you do not study these companies and their approach and start solving every problem on your own, you will end up wasting a ton of time and resources. Why? Because there are companies out there that have studied and already solved these problems. While the context might be different for each of these companies, unless you study the approach adopted by these companies, you will have no point of reference or standard formula to solve your problem in this space.

Whether it is robots going down escalators, or integrating robots with elevators, or finding ways to minimise the interference caused by sunlight on LiDAR sensors, each of these problems has been solved or at least risk-mitigated by other companies. Hence, it is important to study them before you start solving your problem, so you have a series of approaches you can leverage as needed based on the context of your problem.

So, if you are a student about to kickstart your career in robotics, invest your time and resources wisely. Do not fall in love with the robot arm located on your desk. Instead, take a robotics company that has successfully built and deployed robots at scale and deeply study their approach. Speaking of approach, always prioritise approach over technical know-how.

02. Prioritise Approach over Technical Know-how

While most students spend a great deal of time understanding how the Kalman filter works or how the ROS global and local planners solve the autonomous navigation challenge, few take the time to explore the alternative approaches used in the industry to address these same challenges.

A good example here would be that of a Kalman filter. A question that I often ask students is: do you really need a Kalman filter every time you need to fuse data from sensors? Could it be substituted with a deep learning–based fusion model? Have you studied other approaches companies use to fuse this data that are cost-, power-, and compute-efficient?

These are the kinds of questions you begin to ask when you start thinking critically about your approach to solving problems in the robotics world. When you indeed start thinking critically about your approach, it goes without saying that you will soon start realising the importance of developing breadth of expertise. Why? Robotics is a unique field, and to truly solve problems in the real world, you need a lot of techno-operational breadth versus depth.

That brings us to the next pointer on breadth.

03. Prioritise Breadth Over Depth

Most robotics students go miles deep into ROS and other robotics algorithms and frameworks. This, by the way, is not a wrong approach—but it becomes a risky one if you rely solely on robotics to succeed in your career as a robotics engineer. Why?

Just as oxygen is essential to the human body, yet you don't become a doctor by solely studying an oxygen molecule or the respiratory system alone, you have to study other systems and processes that are equally critical to our body with depth and clarity. The same holds true for robotics.

A robot is made up of both robotics components, such as sensors, planners, and controllers, and components that may not seem directly related to robotics but are just as integral, such as the network module, security module, telecom module, tablet, battery type, and docking system. You may ask: why do I need to study these components?

What happens when a robot loses access to Wi-Fi or a 4G network? What if you need to deploy the robot in another country—will you need a different telecom module? What frequency bands do you need to support? What happens when your robot has to go through cybersecurity certifications? And how do you know how secure your robot really is? Should you select LiPo battery or $LiFePO_4$ battery for your robot? Which has better thermal tolerance? The list of questions and edge-case scenarios goes on. You need to be able to analyze and answer these questions to sell a robot in the first place or to even represent a robotics company as an engineer.

You need to study robotics with breadth in mind. Put yourself in the shoes of the customer who is going to buy your robot, and reverse-engineer all the worries they might have—all the areas that could lead to failure or bottlenecks—and study them deeply. Most of these challenges have already been solved by someone out there building robots. That being said, you must study their approach and learn how to adapt that to your own robot or use case.

Always prioritise breadth over depth. The breadth here refers to the breadth of system components that make up a robot, and also the type of robots that are available in the industry, and that leads us to our next pointer, diversification.

04. Diversify – Don't Just Bet Your Career on Mobile Robots

A vast majority of studies coming out of academia today are overly focused on mobile robots, and that is understandable. However, real-world conditions demand solutions that often lie outside the scope of mobile robotics. Try programming a mobile, i.e. wheeled, robot to go up and down an escalator or a staircase; that is a recipe for disaster. Instead, a humanoid or a legged robot, i.e. a quadruped, would perform much better in these scenarios.

This pointer is meant to ensure that you do not limit yourself to a specific type of robot. Instead, diversify your skillset and develop a coherent understanding of all major classes of robots, such as quadrupeds, industrial robot arms, and humanoids, that are deployed in the industry today. We covered some of these in this book, ranging from Spot, the quadruped robot that does a great job in rugged environments, to Stretch, the mobile robot arm explicitly developed for the warehouse and logistics industry, and then the humanoid robots that are being developed by Tesla to automate the manufacturing process.

It is important to study this entire spectrum of robots so you can develop a holistic understanding of the domain, better understand the needs of your customer, and develop solutions that are catered precisely to address those needs. This approach will help you understand both the challenges of operating robots in different environments for these customers, the corresponding constraints and risks involved, and how you can approach them with precision and clarity. That brings us to the next pointer, i.e., solving real-world problems for customers.

05. Customer First, Robot Second

As an engineer or student, constantly put yourself in the shoes of your customers. Whenever you visit a mall, a warehouse, or a factory, try to understand the challenges people there are facing. Take a pen and paper and try to answer these questions: which of those problems could be solved with robots, and which should not be? And why? Of those that can be solved with robots, what should be my approach to validating this problem, collecting requirements from these folks, and building a prototype?

Do not just start building new robots to address those challenges. Instead, adopt a mindset of analysing these real-world scenarios, evaluating the types of robots you could deploy, and documenting the risks, challenges, and potential solutions you would develop. Repeating this exercise across a wide range of scenarios like these will help you truly understand the challenges faced by modern robotics companies and customers. It will also help you build a solution-oriented mindset.

Speaking of challenges faced by robotics companies and customers, while there are many that we can discuss, in the interest of time, one major challenge that often comes up in these discussions is the service or the revenue model.

06. The Economics Matter More Than You Think

It's appalling to see how rare it is to come across a robotics engineer who has done enough homework on this front. What do I mean by that? As a robotics engineer, a vast majority of your time is focused on developing robots, and that, to be honest, may not be the most appropriate use of your time. Why? Most engineers already know how to build a robot from a technical point of view; what they do not know is how to build a robot that addresses all the requirements of the customer and generates revenue for the company while minimising the downsides or risks of deploying a robot at a customer site. That is the problem you want to focus on.

Specifically, there are two of the most important variables in this question that you should keep an eye out for: value for the customer, and the value for the company you work for or build for. These two have to be in sync, and that is extremely hard to achieve in this industry because robots are still seen as desire-based machines, i.e., good to have and not need to have, by most customers. Hence, the price they are willing to pay for a robot-led service does not, in most cases, align with the value offered, and that leads to the death of most robotics start-ups that do not understand this point early on. Do not do that. So, what should you do?

A big part of getting this equation right is ensuring your revenue model is clearly defined from the get-go, and that it factors in any near-term technical, manufacturing, supply-chain, regulatory, and other headwinds that you can see ahead of time. In simple terms, the structure you use to develop a pricing model for your service should be such that customers are willing to pay that price, and you, as an employee, are able to generate reasonable profit for the company at the same time?

That is probably the hardest question to answer on paper as a robotics engineer, but one of the most defining ones for your career. Why? Because it shows that you do not just excel at building robots, but that you excel at building those that create long-term value for both your customer and your start-up. So, where do you start?

To answer these questions, you would need to start studying the existing companies and their pricing models. Specifically, study all aspects of their process: why are they building that robot? How did they determine that the problem is real and worth solving at scale? How did they validate and convey the pricing model to the customers for their robot-led service?

Examine the pricing models being used to generate revenue. Are these companies selling robots as hardware platforms, as hourly subscription services, or as a combination of hardware and service offerings? How are they generating revenue—and which models work best for both the customers and the robotics companies themselves?

This is the homework you should do before you go out there and build or sell robots. This will ensure you are securing the financial lifeline of your company as you build and launch these robots. It is this financial artery that, at the end of the day, will help you create sustained value for both your customer and your employees in the long run.

While these pointers are not designed to provide you with an exhaustive list of all the traps that robotics engineers run into in their career, I hope they serve as guiding principles and help you develop a robust career trajectory in this field. Now, speaking of career trajectory, a lot of students ask for guidelines on landing a job in this field. We will dissect these guidelines into two sections. First, we will cover the traps that you must be aware of if you are looking for a job in this field in the next chapter. Then, in the following chapter, we will cover the approach you can take to identify and secure a meaningful job in this domain.

Chapter 19: The Job Search Traps No One Warns You About

Every year, millions of students earn engineering degrees across the world. India and China alone produce about 1.4 million each, and the U.S. adds another 200,000 or so.[57-59] These figures include graduates with BS/BE and MS/ME degrees. Students who have spent years mastering their domain. Yet a majority of them struggle to find good jobs for months, sometimes years, after graduation. This is especially true in robotics.

So, as a robotics engineer, what are some of the traps you should keep in mind as you approach this job search process? And why is it important to know them ahead of time?

The simple reason is that they're not obvious at first. Even worse, they create an illusion of productivity, so they feel reasonable while you're falling into them. But the cost of staying in one can be enormous. I've watched students lose months—sometimes years—circling the same loops of trial and error, burning through their energy, savings, and confidence. Which is why it's worth naming the biggest ones early. The traps that catch most people not because they're careless, but because they're doing what they genuinely believe they should be doing.

01. Lack of Purpose Sinks Most Dream Boats

The first one on this list is lack of purpose. There's nothing more dangerous in this world than not knowing your *why*. A vast majority of engineering students and even working professionals treat position, salary, location, and the prestige of putting the logo of a Fortune 100 company on their resume as the primary motivation for their job search. This is not a theoretical claim; it is, unfortunately, evident in how students are making decisions these days. Over the last few years, I've spoken with many students and asked this question again and again: why do you want that job? Why are you willing to go through seven rounds of interviews, get rejected, feel crushed—why?

What became glaringly obvious was that most are driven by a lack of purpose—or by the wrong purpose. Yes, you need to earn money to repay loans, pay rent, and buy groceries—but those are the byproducts of pursuing your mission and doing work aligned with it. Most have completely ignored the need to define the mission and are being led by wrong intentions. That is a strategy for disaster. If this is you, it's time to take a step back. It's time to define a long-term mission for your own professional life so you can work on problems in your domain that actually matter to your customers and employers—problems that are hard and require an immense amount of time, energy, focus, and most importantly endurance.

If you are chasing entry-level jobs which means you are trying to solve easy problems, first you're competing with millions in a rat race—and second, you're running in circles. If you keep solving easy problems, you build a reputation to be eligible only for entry-level jobs. Why? Your experience is built on solving easy problems. Be clear about what you want to work on—and why. Your mission statement should deeply define the problem—or tight set of problems—you want to solve, the specific set of industry verticals that are dealing with these problems, and the corresponding technical expertise that you are willing to build and hone continuously to complete this mission.

More on that later. For now, if this is one of your blind spots, please do take time to write it down and think deeply about it.

02. No Strategy = No Direction

Now, among the small group of students who had defined their mission, a majority had failed to define a clear strategy to act on it—that is, to get a job aligned with their mission within a given timeframe. In simple words, they had no structured approach to identify and filter out companies and roles that aligned with their purpose. Some would randomly apply on job portals; others were hopelessly attending webinars, meetups, and networking events, thinking someone would hand them a card with an interview offer. This is another disaster, and I can't emphasize enough how many students told us that networking events were their primary way to "get in front of the queue."

There's nothing wrong with attending a networking event—except when you do attend one without understanding the "why" or the purpose of that event. If you don't, you're simply a potential customer at these events. You don't want to be an unguided missile. Know the difference between being a potential customer at these events versus a potential value creator, and understand which platforms and events are worth your time—and which ones are not. Don't be a customer.

99.9% percent of these events are not meant to help you solve your problem—they're designed to convince you why you should buy a product X or a service Y. They are designed to hijack your most important asset, i.e., your attention, and in simple words, your mind. Beware.

Speaking of attention, just as much amount of focus and attention should be put on crafting the right strategy, an equal amount of it should be put into surgically weeding out bad strategies. This is where the next lesson comes handy.

03. Spray & Pray Outreach = Gambling.

You are an engineer, not a gambler. This is another spectacular blunder that 99.99% of engineering students make. The vast majority of students and engineers say that time is their biggest constraint. Many have education loans to repay or are international students operating under visa-related deadlines. Given this time constraint, most resort to the volume-first, or the gambling approach, i.e. the "spray and pray" method. Many are spending about four to six hours a day on job portals, applying for hundreds of newly posted jobs in the hope of landing an interview. When I asked some of these students, what percentage of these applications actually converted into interviews, the answer was shocking. Any guesses? For those who relied on the spray and pray method, the average was one interview for every 200 applications.

No surprise there. If you are in this boat, you're wasting an entire week—168 hours—to apply for roughly 200 jobs just to secure one interview. Most students did this for three to four months straight before deciding to explore our courses. Now imagine doing that at every stage of your career: once for your entry-level job, then again for mid-level, and once more for senior-level. That's roughly three months per search—almost a full year of your life wasted in this repetitive and soul-crunching job-hunting process. Ask yourself, how much do you value one year of your lifetime?

It is unfathomable how much time engineering students pour into these activities. Do not be that student. Your time is extremely valuable and should be surgically invested to complete your mission. Now, once you start thinking surgically about your investment of time in these activities, and eliminate bad approaches for your job search, the next key parameter you want to keep an eye on is the needs of the market.

04. Your Degree ≠ Market Demand

If a student gets the mission and the approach right, their next challenge usually boils down to misguided beliefs. One of the core ones here is related to the education system. It helps to understand that the education system and the industry operate on different timelines. The industry is usually decades ahead of the education system, if not more. That doesn't mean the education that you get in the academic system is of no value; in fact, it indeed helps you build some parts of the foundation, but not necessarily all of them.

Most students are victims of this hypothesis that has been passed down to us by previous generations—get an engineering, medical, law, or business degree and get a job. From getting a degree to getting a job is not one hop. There is a big delta out here that you need to understand, and that is industry-specific skills and experience. You will need more than that foundational skill set to solve real problems. That brings us to our next pointer, i.e., experience.

05. The Experience Gap Is Real

This is the next challenge most students run into—experience. Lack of real-world experience sinks the remaining 99% of dream boats in the industrial ocean. The demands of the industry are surgically precise and exponentially growing each day. Gone are the days when you could just program in C or Python and apply ROS principles to build a robot. You need real-world experience. Period. To land the right job in engineering today, you need to have a clear understanding of the problems companies in your domain are facing and how you can help them solve those problems.

For example, if you are a CS (Computer Science) engineer applying in the LLM space, you must understand the risks involved in building modern LLMs (Large Language Models) and VLMs (Vision-Language Models): what happens if data leaks, how to handle privacy laws across regions, how to build a revenue model that is justified for both users and the company, and how product choices affect overall revenue and adoption curves. From tech to users to business, you must think from all vantage points and make clear decisions. How do you build this experience?

Just as you can study and master the technical stack of your domain layer by layer, you can develop layers of experience in your domain. You can study how different companies approach a set of problems in your industry and develop a deep understanding of solutions that are widely adopted in your space. This way, if you ever need to help a startup or a company address real-world challenges in your space, you have a set of solutions that you can evaluate and choose from, or build on, based on the context.

The same logic applies to students pursuing careers in robotics and self-driving cars. You must understand the underlying risks inherent in your domain, the deployment challenges involved, the solutions that teams around the world have already tried to address these challenges, and the trade-offs that come with implementing each solution. To make sound decisions, you should be able to study and explain—at minimum—a dozen of these approaches and know when to use which, in which scenario. That is how you hone your decision-making skills and build solutions to address problems in the real world within the context of your domain. That, in fact, is our mission at Boring Sage. That is what we teach in our courses.

In each of our courses, we help students understand the approach adopted by leading industry players to build modern robots and AI models. Specifically, we help students understand what challenges and risks these teams ran into as they started building these products, how they addressed those challenges, what approaches were evaluated by them, and why they chose the solutions they did to solve the underlying problem. Thus, using this approach, we help students build deep professional expertise in their domain—expertise that commands demand and one that sets our students apart in interviews for their jobs.

Before you develop those skills and experience you have to address a few mental traps that often blur the lens of most engineering students. What are these? That's where the next pointer comes in handy.

06. Mindset Trap I – There Are No Jobs in the Robotics & AI Domains

Many students claim that there are no jobs in the robotics and AI space. That is not true. From a logical perspective, a job is a problem in your domain that a company wants someone to solve within a given timeframe. So, if we say there are no jobs, that also implies there are no problems left to be solved in your domain. Now make your rational decision here: is that the case in robotics? Hopefully not yet.

So, based on pure logic, there are jobs. In fact, there are thousands of jobs currently available in both the robotics and AI domains, and less than 10% of these are advertised as ones suitable for recent graduates on job portals. Why? A vast majority of them demand deep expertise and experience—rightfully so.

That does not mean you are not eligible for the remaining 90%. If you are thinking like that, that is a trap. We have helped students address this bottleneck through our courses. That brings us to the second mindset trap.

07. Mindset Trap II – You Do Not Have Experience, and Hence You Will Not Be Eligible for These Jobs

Students tend to think that if they do not have experience that is in line with the role posted by an employer, they will not be eligible for that job. That is, again, a trap. While lack of professional experience is indeed one of the major asks these days in the robotics and the AI space—and rightfully so, given the risk involved in deploying a malfunctioning robot or an LLM model amongst humans—that does not mean you cannot build experience. You can, in fact, do that. We have helped students address this bottleneck. Pointer 05 provides an overview of the approach we taught our students to deal with this trap.

08. Mindset Trap III – You Need to Reach Out to Recruiters and Attend Networking Events

Big mistake here. Yes, you can indeed reach out to recruiters and attend networking events, but make no mistake: you are competing against 1,000 other applicants, and unless you think you have a reason to stand out in this recruiting process that is basically spray-and-pray or in simple words "Gamblers Approach", this may not be the right use of your time. Pointers 02 and 03 go deeper into the dangers of falling into this trap and provide guidelines on how to make clearer, more rational choices about which events are actually worth attending. I can't emphasize how many of our students came to us and told us how much time they wasted until they applied our approach and took our course. You don't have to take our course, but at least craft a targeted strategy for your job search.

09. Mindset Trap IV – Most Companies Do Not Sponsor Visas, So I Am Out of Luck

Many students assume that most companies do not sponsor visas and, therefore, they are not eligible for jobs posted by these companies. This is again, another trap. A job is essentially a problem that needs to be solved. Most startups are well-funded and would happily sponsor you a visa as long as you can solve a problem for them that is worth 10x the price of that visa. So, always raise the bar for the type of problems you are willing to solve for a startup. Ask what the most pressing problem is, surgically develop a series of solutions within 30 days, and propose those to the founders or the appropriate engineering teams. Speaking of solutions, some of you might ask: what about the experience needed to build these solutions, let alone propose one? That brings us to the next pointer.

10. Mindset Trap V – I Don't Have Experience, So I Can't Help Startups Yet

There are many ways to gain experience outside of securing a job. We've had students come to us with backgrounds in ROS and Python who had built ROS-enabled robots yet were still struggling to land a job. These students held BS and MS degrees in robotics—some had even published research papers on improved versions of SLAM—but had little understanding of how different types of robots are actually built and deployed in the real world: quadrupeds, humanoids, space rovers.

More importantly, they didn't understand the real-world challenges involved in building these robots and the corresponding solutions they could use to address them in the field. We've helped them overcome this bottleneck through our courses. Pointer 05 goes deeper into the dangers of falling into this trap and provides an overview of the approach we taught to our students to deal with this trap. That brings us to the last mindset trap.

11. Mindset Trap VI – Apply to as Many Jobs as I Can in Robotics

Before we address this trap, it helps to take a step back and understand that you are an engineer and not a gambler. There is a big delta there. Take time to digest this fact. If you are applying the spray-and-pray or the gambler's approach in your career, you are wasting your valuable time. Pointer 03 goes deeper into the dangers of falling into this trap, while Pointer 05 offers guidelines to help you make a more rational decision if you find yourself in this boat. Instead, find a few companies that are truly aligned to your mission and skill set, and try to solve the problems these companies are facing. Some of you might ask how do you do that?

A Structured Approach – Your GPS for Landing the Right Job

We recommend a structured approach to our students to help them address these challenges. We cover this approach in our courses on Applied Robotics, AI & LLMs, and Self-Driving Cars by distilling each domain into core principles and real-world applications. You'll learn the exact approaches adopted by teams at Tesla, Google, Bloomberg, and NASA—teams that are building modern AI and robotics systems at scale. Once you have mastered these approaches, you will be able to identify a list of companies and the corresponding real problems you can solve for them. Thus, you can approach these companies, explain your understanding of their product and problems and offer the right approaches from the ones that you will study in our course and demonstrate the value you bring to the table as a potential contributor.

In simple words, once you take our course, you won't become an expert overnight, nor is that our intention, however, you will become an engineering professional in true sense. You will build deep professional expertise in your domain and approach these bottlenecks in your career including job search with clarity, confidence, and credibility.

Now, some of you might ask, what exactly is this approach? I will provide an overview of this approach in the next chapter.

Chapter 20: The Boring Sage Method for Landing a Job in Robotics

Getting an engineering degree is relatively easier than securing the right job. Most students struggle to get this right—not because there are no jobs in this domain, but because they misunderstand what a job really is. A job is simply a problem a company wants to solve for its customers. So, if we say there are "no jobs" in robotics, that would imply there are no problems left to solve.

As we all know, that's not the case. If it were, we would already see robots everywhere—on our streets, in our cars, and inside our homes or apartments. That is not the reality yet, which means the opposite is true: this field is full of challenges, and therefore full of jobs. In fact, we're only going to see *more* jobs as companies continue to run into more complex sets of problems and risks that need to be addressed with the growing adoption of robots in our lives.

Speaking of risks and challenges, in the previous chapter we learned the core ones that most students encounter on this journey. Now, in this chapter, you will learn a step-by-step framework, i.e., a structured approach, that you can use to address those risks and secure a job in this field.

This approach is the culmination of a series of conversations I've had with engineering students from around the world over the last few years — from Nepal to Ethiopia, India to the US, and Canada to Italy and France. During each of these mentoring sessions, what became glaringly obvious is that students across the globe run into a remarkably similar set of challenges during the early stages of their careers. A vast majority of these challenges boil down to either their strategy for securing a job or their skill set. These two pillars form the foundation for this approach.

So, after speaking with each of these students, understanding their core bottlenecks, and reflecting deeply on their journeys, I continued refining this approach that I used to guide them. This way, every next student who approached me for advice or guidance benefited from this learning process. Over the years, this cycle of iterated mentoring and research has evolved into a structured framework that I now share with students to help them unblock their path.

Hence, no matter what your current challenge is, if your end goal is to land the right job—the one where you create meaningful value for both the company and yourself—I genuinely hope you find this structured approach helpful in your journey.

Engineering students from around the world who've followed this approach often tell me the same thing: when they took it seriously, followed each of these steps to the tee, and put in the deep work, it actually moved the needle for them. It helped them identify and convert opportunities they once believed were beyond the spectrum of their skill set and experience. We now refer to this approach as the Boring Sage Method.

So, if you're an engineer struggling to land a job, here's how you can start unblocking your own path.

The Boring Sage Method: A Structured Approach to Engineering Job Search

To begin with, it helps to understand that job search, logically, is a five-step process. It all starts with your mission, i.e., the why. Why do you need this job in the first place? How badly do you need it? Then, from there, we define the approach, the requirements, and the execution plan aligned to your mission. We will now cover each of these steps one by one.

Step 01: Define Your Mission — What Do You Want to Work On, and Why?

Step 01 begins with a simple question: *What do you want to work on, and why?* Your "why" must be rooted in your long-term mission, not in financial incentives. Money and titles are outcomes, not drivers. Many students make these external rewards their primary goal, and that is often where their career paths begin to drift away from their mission. The real anchor is your mission, i.e., your purpose. Your purpose will naturally depend on your roots, your journey, and what you hope to build over the long term. But from a professional standpoint, a strong mission is built on the following four core pillars.

1. It solves a meaningful problem in society—using technology that matters to you.

If robotics is your chosen field, your mission should connect you to a real and pressing need in society. It should be deeply meaningful. "Meaningful" here does not mean trendy; it means extremely important for society and extremely hard for the person executing the mission. A problem so hard that working on it forces you to grow as a human being—to elevate your awareness, expand your skill set, and overcome any obstacle, whether technical, financial, or operational, that comes your way.

A strong mission will make you more serious, more committed, and ultimately a better human being. So ensure your mission is strong from the very beginning. That's the first pillar to defining your mission.

2. It allows you to start immediately.

The second pillar you should use to define your mission converges on the importance of embedding a sense of urgency in life. Your mission should not require years of planning before you can take the first step. You should be able to get started on Day 01.

In the context of landing a job in robotics, this means you should be able to start building the expertise you need from the moment you define your mission. This expertise spans both the technical and industry verticals that matter to you. We will define those in the subsequent steps. For now, remember that urgency in taking action toward your mission forms the second pillar of your mission.

3. It has a defined timeline and a clear long-term outcome

The third pillar emphasizes the need to define a clear timeline and outcome for your mission. A strong mission is always defined with extreme clarity. It has a long horizon, a timeline you can truly commit to, and a clear outcome you want to achieve at the end of it. This level of clarity gives your work direction and keeps you anchored when the path ahead becomes challenging.

4. It ensures win-win alignment across both professional and personal fronts.

The fourth pillar emphasizes the importance of creating a true win-win between you, your mission, your professional commitments, and your family responsibilities. Hence, a good mission is not one that drains your health, exhausts your financial lifelines, or negatively impacts the people who depend on you. Instead, it is one that enables you to generate sustainable income, support your family, and fulfill your professional obligations simultaneously, all with peace of mind.

Always remember, long-term stability and balance across all fronts are the hallmarks of a well-calibrated mission. It creates win-win outcomes for you, your health, your family, your team, and your employer. These four pillars together form the foundation you should rely on when defining your mission.

So, what does a well-defined mission look like in practice?

Example of Mission Statement

Here is an example:

"Over the next 10 years, I aim to build a humanoid robot capable of performing everyday household tasks, with a primary focus on supporting aging adults who cannot afford human assistance. My mission is to make this kind of robotic assistant truly accessible and affordable, beginning with the development of a simplified, cost-

effective prototype that I can build at home, and progressively evolving toward production-grade solutions by working with startups engaged in this domain. I aim to do so while maintaining a strong balance between my personal and professional commitments, and at the same time creating long-term value across both."

Decoding the Mission Statement

This mission works because it addresses a deeply meaningful and underserved problem in society by focusing on aging adults who lack access to affordable assistance, thereby fulfilling the first pillar of impact and significance. It also establishes a 10-year timeline and a clear outcome, offering structure, direction, and a long-term horizon to work toward, which aligns with the third pillar.

More importantly, this long-term horizon reinforces the fourth pillar of win-win alignment. A 10-year timeline encourages patience, discipline, and sustained focus, allowing the individual to build value at a steady pace without compromising health, financial stability, family life, or professional well-being. This long-term mindset supports a sustainable approach that balances professional and personal commitments, ensuring stability across all fronts.

At the same time, the nature of this mission encourages immediate action. It enables the individual to begin building relevant expertise from Day 01 through research and prototype development, reinforcing the pillar of urgency and execution.

When defining your own mission, look for areas in society that are extremely underserved and where people are urgently waiting for solutions for years. This is where you can connect your mission with your skill set and create long-term value for society in the truest sense.

Now that you have your mission defined, the next step is to define your approach. Assuming you've completed your engineering degree, or are on the path to doing so, this approach focuses on your ability to identify the right opportunities that will help you build the experience you need. In other words, the professional roles that help you inch one step closer to your mission.

Step 02: Develop A Deep Understanding of Your Core Primary & Secondary Skills

Start by reflecting on the top two technical areas of expertise that are important in your domain and that you are genuinely good at, compared to all the other technical skills listed on your resume. For example, if you are a software engineer, your primary strength could be backend development in Python and your secondary strength could be AWS cloud. In this case, backend and cloud are your primary and secondary core skills.

The same principle applies in robotics. Some of you may find that your primary strength lies in developing high-performance perception systems, helping robots develop a strong understanding of their surroundings, while your secondary strength could be backend software development in Python. Others, especially those coming from a mechanical engineering background, may discover that their primary strength lies in developing accurate 3D models of robot parts, with a secondary focus on building high-performance battery and docking systems.

For those who come from an electrical engineering background, your primary strength might lie in designing industry-grade PCBs to integrate sensors into a robot's architecture, while your secondary skill could focus on sensor testing and validation. No matter where your background sits, the goal here is simple: clearly identify your primary and secondary strengths.

Now that you've finalized your mission and skill set, the next step is to narrow down the industry verticals that align with both.

Step 03: Shortlist Industry Verticals Aligned with Your Experience

Whether you already have experience working on academic projects, internships, research projects, real-world projects, or even hobby projects, list the industries where you have the most exposure. As long as that experience falls within the robotics spectrum and helps address real problems for a given industry vertical such as healthcare, insurance, retail, ecommerce, warehouse logistics, or aerospace, it counts.

For those who are just getting started, go online, read books, take courses, and begin developing that experience. You don't always need a job to build industry-specific experience. You can study, research, and build your understanding of the industry vertical you are genuinely passionate about.

The goal here is not to become a master of that industry vertical. The goal is to develop *enough* understanding so that you can speak about that industry, its challenges, risks, and opportunities, with clarity and confidence. This means you can understand the customer's problems in that space inside out, and speak about them confidently in your interviews.

Step 04: Define Your Core Inputs

By this time, you will have developed four key inputs: your mission, your primary skill, your secondary skill, and your industry vertical. These four inputs are critical for shortlisting companies and roles—in other words, identifying opportunities that are worth your time and where you can create value for both the company and yourself.

How do you do that? That's where the next step becomes useful.

Step 05: Define the Five Criteria to Shortlist Companies and Roles

Once you have defined your mission, your skills, and your industry vertical, the next step is to identify which companies and roles are truly worth your time. This is especially important in the robotics space, where the mortality rate of startups is extremely high, and where not every opportunity that comes your way will help you inch closer to your mission.

So how do you save your time and energy here? How do you minimize the risks associated with working at an early-stage company? You do so by applying the following five criteria as your minimum filter before shortlisting any company or role.

Criterion 1: Alignment with Your Mission, Skills, and Industry Vertical

No matter which company you choose, its work, i.e., its mission, should align directly with your own, and the corresponding role should allow you to apply your primary and secondary skills within your chosen industry vertical. This ensures that every opportunity you pursue keeps you on the path toward your long-term mission.

Why is this important? It takes weeks and months to develop a laser focus on your mission, and just one wrong opportunity to destroy that momentum and focus. So be extremely careful with every step you take on this path. The first and most critical one is the company or project you choose to take on. It should be surgically aligned with your mission and your skill set. Once you identify such a company, the next step is to ensure it has a strong financial lifeline.

Criterion 2: Funding Status and Customer Adoption

If the company is a startup, make sure it is adequately funded and has at least a handful of real customers generating revenue. Most companies will not reveal these financial details openly, so a good proxy is to study the trajectory of founders and the investors. Reputable founders and credible investors, especially those with a history of successfully executing extremely hard projects, are strong signals of long-term stability.

As you study these trajectories, look for long-term patterns that signal endurance, persistence, and hard work, patterns that have consistently yielded strong results. Pay close attention to the outcomes of the projects they have been involved in. Ideally, you should see a mix of both failure and success. A patter of success that has emerged after a series of failures is a strong indicator of resilience and solid execution capability. In summary, these trajectories should include patterns indicative of both success and failure, not just one or the other.

Remember, robotics startups are far riskier than typical software startups. This makes this criterion significantly more important here than in many other industries.

Criterion 3: Is the Problem Real, Painful, and Urgent?

A good robotics company must be solving a real, painful, and urgent problem. One useful indicator is whether customers are willing to pay meaningful fees *today* to automate the solution for this problem. If the problem is trivial or "nice to have," the company is unlikely to survive long term. Look for companies addressing challenges that are extremely painful for customers and hence demand urgent solutions. This ensures that the work you do carries significant weight to it. How do you do that?

You have to make an internal evaluation here. You need to study that problem independently and answer this question for yourself objectively. Can you write down eight *quantified* reasons why this problem is real, urgent, and one that customers are willing to pay for? Can you list those reasons in a way that cites evidence from media reports or other credible sources? Can you speak directly with these customers? You need to do your own research to clearly separate companies driven by needs from those driven by desires here.

Criterion 4: Company Culture and Team Alignment

Company culture matters more than most engineers realize. The difficult part, however, is evaluating that culture *before* you step into the company. So how do you do that in a meaningful way?

One way is by observing how team members speak with you during interviews. Pay attention to whether these conversations reflect curiosity, long-term thinking, technical depth, mutual respect, and a genuine interest in doing deeply meaningful work. Culture shows up in the way people think, the questions they ask, and the values they emphasize without being prompted. If only the founders talk about purpose, learning, and growth while the rest of the team seems disengaged, or the opposite, where the team cares deeply but the founders do not, that is a clear red flag. Strong cultures are consistent. You should notice similar attitudes, energy, and urgency across multiple conversations, not behavior isolated to one or two individuals.

A healthy culture leaves you with a sense of coherence and alignment. A weak one leaves you confused or uncertain. That feeling is often your most accurate signal, so don't ignore it. In simple words, an interview is an opportunity for both stakeholders, you and the team, to interview each other and get to know each other. It's never a one-way process.

Criterion 5: Growth Trajectory and Long-Term Stability

Study the company's overall trajectory. Look at their product announcements, presentations, hiring trends, and any publicly available information related to customer footprint and revenue trajectory. Companies that demonstrate strong and sustained momentum across these fronts are more likely to retain employees over longer periods and reflect stronger long-term stability. This criterion is especially important if you are targeting robotics startups. Why is it so important?

Product development cycles in robotics are extremely time- and resource-intensive, which significantly increases operational risk and directly impacts a company's ability to survive, let alone thrive. As a result, evaluating operational stability here is not optional; it is essential.

Step 06: Eliminate Bad Outreach Strategies

Now that you have finalized your list of companies and roles, you need a clear approach for outreach. Most students make a spectacular blunder here by applying for hundreds of jobs on third-party portals. You should use these platforms to discover companies that may need your skills, but not necessarily as your primary channel to apply to those jobs. If you solely rely on these portals, you are competing with thousands, sometimes millions, of other applicants. The odds are stacked against you.

So how do you solve this challenge?

Step 07: Define a Structured Outreach Strategy and Land Interviews

There are many ways to address the outreach challenge, but the one approach that has consistently worked for all our students is outlined here. It's a structured approach that enforces the need to slow down, study a few high-quality startup teams, understand their product and the problem they are trying to solve, and honestly assess whether you have the skills to contribute to those challenges. Once that level of clarity is established in your mind, you then reach out to those startups through either 1:1 DMs or emails. This approach requires value-enabled depth, not volume-enabled breadth

That naturally leads to this question: how can you learn more about a startup and its product portfolio? How do you study a company deeply enough that you can truly understand its core bottlenecks, the areas where you can contribute and create value?

Step 08: Surgically Evaluate Startups — Their Product & Core Bottlenecks

While there are several approaches you can take to evaluate a startup, it's important to start with your core inputs: your mission, your primary and secondary skills, your industry vertical, and the five criteria. Use these to target robotics startups that satisfy these filters. Do this with almost surgical precision.

Study their products and challenges using publicly available information: company websites, engineering blogs, founders' posts, press releases, webinars, demo videos, investor presentations, and open-source repositories, where applicable.

Narrow down two or three problems that you can reasonably confirm are important for the company and that match your skill set. These should be problems you can speak about clearly and propose solutions for, based on your technical background and understanding of the industry, both before and during your interviews with these teams.

Step 09: Apply the Outreach Flow

Now use the outreach flow explained below (refer to the Outreach Flow section) to make contact with founders and the teams leading these startups. Focus on clearly demonstrating your understanding of the company's challenges and offering a well-articulated, thoughtful solution.

Send this message to the founder, CTO, engineering manager, or hiring lead via email or direct message, on the social platform where they are most reachable. This process and the exact structure of this outreach flow will be covered in detail later in this chapter.

Step 10: Be Respectful & Play the Long Game

As you conduct this outreach, always lead with kindness, mutual respect, and patience, regardless of the outcome you may have to digest. In the robotics world, a "no" rarely reflects your capability or potential. It almost always means "not now." These companies operate under extreme pressure, scarce resources, and tight deadlines, and their decision timelines are rarely linear. In such environments, persistence, when practised with professionalism, becomes the defining factor that separates serious candidates from the rest and builds long-term credibility.

It is also important to remember that robotics companies operate on an exceptionally long horizon. Products take years to design, build, test, and scale, unlike many purely software-enabled services. As a result, even a "not now" at this stage is not a loss; it is often a quiet win. It signals your ability to build long-term relationships, absorb rejection with maturity, and remain committed to your mission. In many cases, startups do return to candidates whose outreach stands out as thoughtful, value-driven, and serious. When your message reflects a genuine understanding of their problem and a clear sense of how you can contribute, it leaves an impression that extends well beyond the immediate hiring cycle. This is what playing the long-term game truly means.

Step 11: Calls Will Follow

If you follow this structured process with absolute commitment and a deeply serious attitude, calls will follow. You will eventually speak with founders or hiring teams who recognize the value you bring to the table. From that point onward, conversion depends on your ability to understand the depth and breadth of their challenges, i.e., bottlenecks, evaluate and demonstrate how you assess different approaches to address those, and then your ability to articulate solutions and propose the same.

That is essentially the approach we teach in our courses at Boring Sage. Students learn not just the core principles in robotics, but also how companies around the world are building modern robots, how they evaluate risks, different approaches to address those risks, and build solutions to bring these robots to life. Our goal is to help you develop a thorough understanding of your domain, both on the technical and industry vertical fronts, so you can build deep professional expertise. Expertise that companies demand, and that will set you apart in your interview and at your job.

Now that we have covered the 11 steps you can use to land interviews, the next question is: what does the outreach flow referred to in Step 09 actually look like? We will now cover that in detail.

Outreach Flow

The outreach flow that I recommend to students here at Boring Sage consists of three components:

1. **Your five core inputs**
   Mission, primary skill, secondary skill, industry expertise, and startup evaluation criteria.
2. **Startup portals**
   Curated platforms to identify, assess, and filter robotics companies that align with your direction.
3. **A structured message flow**
   A clear, value-driven approach to conduct direct outreach with precision and intent.

We covered the first component here, i.e., the five core inputs, in the 11 steps discussed earlier in this chapter. Now we will focus on the next two components, the startup portals and the structured message flow, that you can use in your journey.

Part A: Startup Directories

While there are many portals available online to help you scan for robotics startups, deep-tech teams, and early-stage companies working in this area, here are a few that have helped students identify and filter companies and roles aligned with their mission. These portals should be used primarily for discovery and research, not as your sole application channel.

1. Y Combinator Directory of Startups

A comprehensive directory of Y Combinator-backed startups across multiple sectors, including robotics and deep tech.

Link: https://www.ycombinator.com/companies

2. Work at a Startup (YC)

YC's dedicated job platform that allows candidates to apply directly to thousands of YC-backed startups globally.

Link: https://www.workatastartup.com

3. Startup Gallery

Features a curated collection of VC-backed startups, useful for discovering emerging companies and exploring early-stage teams.

Link: https://startups.gallery/

4. Landing Club

Lists open roles across startups and provides additional information such as visa sponsorship for each job posting.

Link: https://landing.club/stamplist

5. Sequoia Capital Jobs

A curated job board featuring roles across Sequoia-backed portfolio companies, including engineering, design, sales, and operations.

Link: https://jobs.sequoiacap.com/jobs

6. a16z Talent (Andreessen Horowitz)

Andreessen Horowitz's talent network and job board designed to match candidates with open roles in its portfolio startups.

Link: https://a16z.com/jobs

7. Lightspeed Venture Partners

Lists open roles at Lightspeed-backed startups worldwide, particularly useful for identifying high-growth deep-tech companies.

Link: https://jobs.lsvp.com/jobs

How to Use These Portals Effectively

These directories are not meant to be used as high-volume application engines. Their real value lies in helping you discover, evaluate, and shortlist startups that align with your mission and skill set.

Use these platforms to identify companies that are actively building in areas relevant to your domain. Study their product pages, team composition, funding background, and recent activity. Look for signals that reflect alignment with your interests, technical strengths, and the kind of problems you want to solve.

Once you identify promising companies, shift your focus away from simply clicking "Apply". Instead, move towards direct, thoughtful outreach that we discussed in the previous 11 steps. Whenever possible, reach out to founders or engineering leaders directly before contacting recruiters. Robotics founders especially appreciate initiative, technical depth, and a serious problem-solving mindset. Research the team, understand their bottlenecks, and approach them with a well-rounded message that is grounded in the value that you want to offer.

In short, treat these portals as a discovery and intelligence layer, not as a shortcut to mass applications. The goal is precision and value, not speed.

Part B: Outreach Message Templates

Below are example message flows for robotics students. The goal is to connect your skills directly to the startup's real challenges and position yourself as someone who deeply understands their problems. In simple words, your goal is to save their time by doing your homework and delivering value by clearly articulating their problem or core bottleneck and proposing solutions that align with it, demonstrating that you are here to solve real problems and create value from the very beginning.

Message 1: Robotics (Perception / Controls Student)

Principle: Introduce yourself + Demonstrate value

Hi [Founder's Name],

I'm [Your Name], a robotics student with practical experience in ROS, perception systems, and motion planning. I noticed [Startup Name] is building [specific robotics system: AMRs, humanoids, warehouse robots, robotic manipulators].

At scale, challenges like [reliable SLAM, multi-sensor fusion, real-time obstacle avoidance, safety in human-robot interaction] become critical.

In my recent project, I [built a ROS2-based navigation stack / implemented LiDAR-camera fusion / improved planning efficiency by 30%], which addressed similar challenges.

I'd love to explore how I could apply similar solutions to one of your current bottlenecks. Would you be open to a quick call?

Message 2: Light Follow-Up (3–4 days later)

Hi [Founder's Name],

Following up briefly — I was reading about your recent update on [specific product or feature]. One challenge that stood out was [Y], which closely aligns with my experience in [Z].

I've drafted a short outline of how I might approach solving this. If helpful, I'd love to jump on a quick call to discuss further.

(Attach a 1-page solution outline.)

Message 3: Polite Close (1–2 weeks later)

Hi [Founder's Name],

I know you're busy, so I'll keep this short. If now isn't the right time, no worries — I'll continue following [Startup Name]'s progress. If a need arises for robotics or perception support, I'd be glad to contribute. You can reach me at [email]. Thanks again for your time.

Why This Works

- **Message 1** demonstrates a solution-oriented mindset, technical depth, and clear alignment with the startup's real challenges.
- **Message 2** reinforces credibility, thoughtful research, strategic insight, and proactive initiative.
- **Message 3** closes the interaction respectfully while preserving the relationship and keeping the door open for future engagement.

What Students Had to Say

Some students succeed in the first batch of ten startups they target; others need two or three batches of ten startups each. Be patient, improvise, and learn from every outreach. The core principle remains the same: instead of waiting to be interviewed, proactively do the homework. Understand the company, anticipate its challenges, and clearly communicate how you would solve them.

The Boring Sage Method is built to help robotics students break through the noise, develop a deep understanding of the robotics industry, and convert opportunities that once felt out of reach. This approach has helped thousands of students land interviews, build confidence, and secure roles across robotics and AI teams worldwide.

Now that we've covered the outreach approach you can use to secure interviews, the next bottleneck that a lot of students run into is converting that interview opportunity into an offer.

Converting Interview into Offer

Once you secure the interview call, your next true bottleneck lies in converting that opportunity into a definitive offer. This is where a lot of students struggle, not because they lack skills, but because they lack experience. Real-world experience.

Robotics is an unforgiving field; the mortality rate for robotics companies is probably ten times higher than any other tech domain out there. This also means the experience you need to get a good job is equally higher. So how do you build that? That's why we've developed courses. You can explore these at ***BoringSage.com***.

From Mars Rover to humanoids, you'll learn the exact approach adopted by engineering teams to bring these robots to life. More importantly you will develop deep professional expertise in robotics. Expertise that companies demand, and the ones that will set you apart in the interviews and at your job.

Now that we have covered the end-to-end approach you can adopt to build real-world experience, land interviews, and convert those opportunities into offers, some of you may be thinking about starting your own robotics venture at some point. Others may already be on that path and are looking to better understand what to watch out for, the risks involved, and the challenges you will face. We will cover these in the next few chapters.

Chapter 21: Fundraising and Finance

Hard Lessons on Raising Capital and Securing Financial Runway

As I started writing this chapter, I struggled to find the right words to describe what this journey of building a robotics startup truly feels like. How should one even describe this journey? The amplitude of emotions you will experience on this path is much higher than what you might feel if you were building a software startup. The extreme highs and lows that you will witness are hard to put into words. Perhaps the most honest and simplest way to describe this journey is that it is extraordinarily hard. Once you are on this path, extreme pain becomes your recurring revenue, and endurance is your only true companion. And even that feels like an understatement.

Why is that the case? If you are building a robotics startup, you are not building one company. You are building many at once. There is hardware and there is software. There is PCB design, sensor integration, field testing, supply chains, battery systems, firmware, networking, and deployment. Each of these would be a serious business on its own. Hence, if you are building a robotics startup, you are essentially attempting to run all of them in parallel, with extremely limited resources — time, capital, and talent.

On this journey, the real challenge is not just building robots that can thrive in the physical world, i.e., the technology itself. It is really about building them without straining your most vital resources. If you are not careful, this path can quietly erode your physical stamina, your judgment, your financial runway, and most importantly, the morale of your team long before the product ever reaches maturity.

That naturally raises the question: how can founders identify and proactively address some of these core risks and traps that are inherent on this path, and build robotics companies that truly thrive? While we have already explored many of the traps that robotics teams fall into in earlier chapters of this book, we will now focus on lessons explicitly distilled for founders. These lessons are based on a close study of the trajectory of numerous robotics startups that I have conducted over the past few years in preparation for this book.

They are intended to offer a clear map of the core risks and traps that robotics founders must keep a close eye on and navigate proactively as they bring the next generation of robots to life on this planet. You may ask: what is the difference between risk and trap in this context? Risks refer to challenges that are inherent in this path and, in many cases, must be managed and cannot be fully eliminated or addressed. Traps, however, can be identified and proactively avoided.

These risks and traps have been distilled into lessons below, each equipped with actionable takeaways for you, as a founder, to identify, address, and overcome proactively. To make it easier for you to digest each of these lessons, they are organised as follows:

- **Fundraising & Finance:** Raising Capital & Securing Financial Runway
- **Engineering & Talent:** Building High-Performance Products & Teams
- **Sales & Marketing:** Delivering Value & Building Customer Trust
- **Operations:** Building a Robust Execution Engine

We will now cover lessons in each of these areas, one by one.

Fundraising & Finance: Raising Capital & Securing Financial Runway

Lesson 1: Design Your Capital Strategy Around Robotics, Not SaaS

As a robotics founder, capital is both the blood and the oxygen of your startup. Unlike SaaS companies that may need only a fraction of what you will require in each funding round, robotics startups operate on a completely different financial scale. In most cases, you are going to need to raise at least 10x the capital compared to a SaaS startup.

This is where a lot of founders in this space make a brutal mistake: they try to run a company like a SaaS startup and engage with investors with the same mindset, expectations, and performance profile as SaaS founders. Do not make that mistake, because if you do, it will negatively influence almost every downstream decision you make as a founder.

In this world of robotics startups, it is entirely normal to burn through three to four generations of hardware iteration cycles, each consuming 12–18 months, before establishing a sustainable revenue trajectory. Robotics companies that have thrived through these cycles are a proof of this principle. Miso Robotics is nearly 10 years into its journey and is only now scaling the third generation of its robot, Flippy, across live commercial kitchens in the US. RISE Robotics has accumulated 25 patents just to transition from research phase to mass productization phase for what is now the world's strongest robotic arm, capable of lifting 50,000 pounds. Waymo, meanwhile, has been refining its autonomous driving systems for nearly 20 years to reach a point where its vehicles now operate with accident rates lower than those of human drivers.

These examples reflect the fact that your funding strategy must acknowledge this reality, one that is uniquely inherent to robotics startups: long, extremely capital-intensive product development and validation cycles, followed by a late-stage, extreme adoption phase. This product maturity cycle in robotics is fundamentally different from what you see in the software space, where a steady 5–15% year-over-year growth curve is the norm for investors. If you pretend you are a SaaS company when you are not, you will attract the wrong investors, set the wrong expectations, and exhaust your financial runway long before reaching the payoff your product truly deserves.

So how should a founder design a capital strategy in an arena defined by such long product maturity cycles and a heavy burn rate? That is exactly where the next lesson becomes critical.

Lesson 2: Build a Hybrid Capital Stack, not a One-Dimensional Strategy

Robotics founders who thrive and scale their companies do not rely on a single source of capital; they build a hybrid stack that combines retail, institutional, and strategic investors. You may ask why this is the case.

There are two core reasons. First, robotics is an inherently resource-intensive domain, which means you will require ultra-strong financial lifelines — deep pockets, to say the least. Knowing this, if you secure capital only from institutional investors or VCs, you introduce a dangerous single point of failure into your startup. Second, you ignore the distinct value that each class of investor brings to the table — value that no single group can provide on its own. You may ask: what is this unique value that each of these groups provide?

Each of these investor groups add a unique value to your startup that exceeds far beyond capital. If you think about it carefully, institutional investors, i.e., VCs, bring scale and therefore speed. Strategic partners enable market access, i.e., reach. Retail investors enable brand velocity. This hybrid structure ensures your capital lifeline remains resilient in the long run and, at the same time, prevents single points of failure. Companies like Miso Robotics, Knightscope, and Fieldwork Robotics are clear, living examples of this principle.

Miso Robotics operates with over 40,000 retail investors alongside institutional investors and its strategic partner, Ecolab, a $70B public company. It has raised over US $104 million, creating a diversified capital ecosystem that enables long-term financial and operational resilience. Another leading player in this space, Knightscope, Inc., a security robotics company, raised over US $90 million from more than 28,000 retail investors through a crowdfunding process before going public, in addition to securing institutional and strategic funding, demonstrating the power of this hybrid capital structure.[60-61]

Similarly, RISE Robotics blends financial support from institutional investors such as The Engine with strategic partners like Gates Corporation and a growing base of over 1,100 retail investors, who have collectively committed more than $2.5 million.[60] Fieldwork Robotics raised over £1.5 million in a funding round led by Elbow Beach Capital and also opened a Seedrs retail crowdfunding round, with 131 investors contributing approximately £282,760 toward a £390,000 target, demonstrating how robotics companies effectively combine strategic and retail capital sources.[62-63]

This hybrid structure allows founders to maintain strategic control, minimize dependency on any single capital gatekeeper, and activate complementary value streams in parallel: execution speed from venture capital, market access and credibility from strategic partners, and sustained momentum from community investors.

On the other hand, as a robotics founder, if you approach capital with a one-dimensional strategy, i.e., relying on a single class of investors, you expose your company to significant downside risk. This is because a vast majority of investors, even when they understand the difference between hardware and software startups, will still evaluate your performance through a SaaS lens.

The expectations they set — around growth, revenue velocity, and efficiency — rarely align with the realities of the robotics world. As a result, by the time you reach Series B or C, you are already under disproportionate scrutiny and increasingly vulnerable to being replaced or deprioritized in their portfolio. A hybrid strategy, by contrast, gives you extremely high operational longevity, a requirement that is critically important to thrive in this arena.

Some may question this approach. While the role of venture capital and strategic investors is generally well understood, far fewer founders truly grasp the value of retail investors in this space. This naturally raises an important question: how do retail investors meaningfully contribute to your value creation runway as a robotics company? This is where the next lesson becomes especially relevant.

Lesson 3: Retail Investors Are a Force Multiplier

As a robotics founder, it is important to understand that when people invest their own money in your startup, even in small amounts, they develop a strong sense of ownership. They spread the word. They defend your product and service. And when enough of these people exist, i.e., when you have a large group of retail investors, your company gains a distributed layer of advocacy — a belief in your vision at a scale that no traditional marketing campaign can replicate. This belief system travels across countries, cities, and communities, far beyond the reach of any marketing campaign. This is what retail investors really represent. They are not just capital; they are a decentralised belief system operating at a galactic scale.

Miso Robotics demonstrates this clearly. With over 40,000 retail investors, Flippy has achieved brand recognition that stretches into towns and regions where no paid campaign has ever targeted. People know Flippy because someone they know believes in it. In many communities, the names "Flippy" and "French fries" are beginning to feel almost synonymous.[60-61] The same principle applies to Knightscope — the security robot now seen patrolling subway stations in New York, Times Square, and multiple U.S. cities. This is brand penetration at scale, achieved with maximum resource efficiency and a strong focus on community-led growth.

Similarly, RISE Robotics, though in the early stages of its journey, is already seeing its 1,100 retail investors create tangible business impact. These investors are organically opening doors to original equipment manufacturers (OEMs), the companies that design and produce hardware at scale, while also facilitating introductions to trade show managers, industry buyers, and strategic business partners. These introductions and partnerships would typically require large sales teams or extensive business development efforts, yet they are instead being driven by community advocacy channels.

These are not marketing tricks. They are natural outcomes of community-led ownership psychology. These examples demonstrate that when planned and executed strategically, community-backed capital acquisition channels translate directly into brand advocacy and credibility at scale. Put simply, when thoughtfully nurtured, retail investors do not merely support your company; they become one of the most powerful decentralized growth engines your robotics company will ever be blessed with.

Speaking of retail investors, some of you may ask: what does it truly take to launch and execute a successful crowdfunding campaign? What does it take to build an organic layer of trust with retail investors participating in these campaigns? This is where the next lesson becomes especially valuable.

Lesson 4: Crowdfunding Works Only When Brand Precedes Capital

A successful crowdfunding strategy extends well beyond a well-crafted campaign page. Gone are the days when founders could launch a visually crisp crowdfunding page, supported by high-quality media assets and a compelling narrative, and expect capital to flow in effortlessly. That era is long over. Today, crowdfunding works only when a strong layer of belief in your product has already been established in the community — when a brand has been built with an extremely high degree of consistency , and trust has been earned at scale before any request for funding is made. This naturally raises the question: why is that the case?

The reason being, capital does not create trust; it follows trust. And in robotics, that trust must be earned by demonstrating clarity in mission and consistency in execution over time. Your brand is not shaped by what you claim in a moment, but by what you repeatedly demonstrate over a long period of time. We are talking half a dozen years here. Each prototype shipped, each milestone achieved, and each system deployed at a customer site reinforces belief in your trajectory. Only when this belief is firmly established over a sustained period of time does crowdfunding become a viable route to raise funds for your startup.

Miso's journey reflects this reality. Its success is rooted in crafting a deeply intentional brand story centered on "modernizing the kitchen," embodied by the instantly recognisable symbol of Flippy — the robot arm cooking fries. RISE Robotics followed a similar path by identifying its most powerful differentiator — the world's strongest robotic arm — and amplifying it through Guinness World Record validation to create global visibility.[60] These brands were not built through funding; funding followed their credibility. Crowdfunding, therefore, is not a tactical fundraising shortcut. It is a startup discipline tied to clarity of mission and demonstrated through consistency in execution over a sustained period of time.

If your story cannot be understood by a retail investor, many of whom are non-engineers, within five seconds, your campaign will fail long before skepticism even enters the equation. This is why your brand narrative, and your actions must be communicated with absolute clarity, especially to retail investors. Simplicity, coherence, and consistency are not optional. They are fundamental to earning trust at scale. That brings us to the next lesson.

Lesson 5: The Crowd Must Be Treated Like Serious Investors

Retail investors today are way more sophisticated and understand the importance of accurate brand valuation and execution transparency. They are not naïve. Companies that manage crowdfunding campaigns effectively commit to structured communication channels, regulatory transparency, and operational clarity through regular investor updates, quarterly briefings, and publicly accessible disclosures. They understand one critical reality: if you raise capital from the crowd and do not deliver on those expectations in time, or worse, disappear, your credibility is permanently damaged.

Trust is built through consistency. Each communication cycle builds trust, and trust compounds into future capital. The crowd expects clarity, consistency, and competence. If you cannot maintain a disciplined investor communication rhythm, you are not prepared for modern crowdfunding campaigns, regardless of how technically advanced your product may be. Hence, treat retail investors with the same level of respect and urgency as you would treat any other institutional investor. Consistency and clarity in both communication and execution should be your guiding principles on this front. That brings us to the next pointer: at what stage in your startup's journey should you launch your crowdfunding campaign?

Lesson 6: Crowdfunding Is Not a Last-Resort Lifeline

One of the most dangerous myths founders believe is that a crowdfunding campaign can be launched at the eleventh hour to rescue a collapsing runway. In reality, most successful rounds require months of preparation, traction signaling, and brand maturity, with the bulk of capital typically arriving toward the end of the campaign cycle. This makes timing a critical determinant of success, as clearly demonstrated by the path followed by companies in this domain that have executed high-performing campaigns.

RISE Robotics adopted a disciplined "test the waters" approach, securing $800,000 in its early phase before scaling its crowdfunding campaign further. Similarly, Miso approached crowdfunding campaigns as a long-term strategic operation, not a short-term desperate fundraising effort.[60] Knightscope followed a similar path, structuring its crowdfunding efforts as a carefully planned long-term capital strategy rather than an emergency measure, building momentum across multiple rounds long before its public listing. These companies understood a fundamental truth: if you initiate a crowdfunding campaign when your bank balance is already critical, you are designing failure at a structural level. Crowdfunding rewards strategic foresight over crisis-driven fundraising operations.

Speaking of capital, now that we have covered a few lessons around the strategy required to enable financial resilience, the other side of this equation is the cost of raising capital. You must manage this cost with extreme precision. Mismanaging this cost can quietly destabilize even the most well-designed funding strategy. The next lesson explains why this matters and where most founders get it dangerously wrong.

Lesson 7: Manage Cost of Capital Like a Systems Engineer

Crowdfunding is both an art and a science. Most founders treat it as just an art, and that is a costly mistake. If you are not disciplined about every dollar spent to raise capital, the process itself will erode the very runway it is meant to extend. At its core, you are translating attention into capital, which means every dollar spent to communicate your mission, the problem you are solving, and your solution must justify its return. So, how should one manage this cost of raising capital?

Successful founders treat capital acquisition like performance engineering and continuously measure three key metrics: return on ad spend (ROAS), cost per investor, and conversion rate. If you do not see these numbers improving, your system is broken, regardless of how compelling your story sounds. So how should you approach these metrics? What ROAS should you consider healthy at different stages of a campaign? At what point does cost per investor become unsustainable?

While there are no universal standards, historical performance across successful robotics crowdfunding campaigns offers practical benchmarks: aim for a minimum 2.5x–3x ROAS, keep total fundraising costs within 15–20% of capital raised, and maintain a cost per retail investor in the range of $40–$120, depending on the channel used to acquire them. This naturally raises the next question: how should this cost be distributed across different marketing channels such as paid advertising, email, organic content, and community-driven efforts?

One way to think about paid advertising is that it should function as an accelerator, not the primary engine. It must be supported by lower-cost channels such as email campaigns, organic content, investor webinars, PR exposure, and referral loops. As you leverage each of these channels, you must review performance data on a weekly basis and reallocate budget decisively toward the best performing strategies. If you ignore these metrics, you aren't raising capital. You're just burning it.

While these are the core lessons to keep in mind as you raise capital and secure the financial runway for your startup, capital itself is only the starting point. The more critical question that naturally follows is this: how do you deploy that capital to build a product that is extremely valuable for customers and a team that can sustain it? This brings us to the next class of lessons on product excellence and talent allocation. We will cover these in the next chapter.

Chapter 22: Engineering and Talent

Hard Lessons on Building High-Performance Product and Team

While there are numerous ways to think about how you would want to design your team and product, if you study the trajectory of teams that have successfully built robot-led services or products over the last decade, a few core lessons consistently resurface throughout this journey. To make these easier to digest, they have been classified into two buckets — one focused on what founders should prioritize when building products, and the second on what they should prioritize as they build teams to sustain these products. We will now cover each of these, one by one.

Product Excellence

Lesson 1: Design for Failure, Not for Ideal Conditions

If you are building a robot-led product or service, it is safe to say that failures are inevitable. Robots will fail. Sensors will need to be recalibrated, and some might malfunction at the moment you least expect. Cameras might freeze, and no matter how powerful the underlying AI model is, the obstacle avoidance system will fail. Robots may lose access to Wi-Fi or 4G networks and will run into scenarios you never anticipated. While you cannot prevent each of these edge cases, as the list is almost infinite, it helps to understand that your product design philosophy should start by assuming these failures are inevitable. Hence, your number one priority from a product design perspective is to ensure there are safety mechanisms designed in place, so your robot recovers gracefully from each of these edge cases without causing harm to any person, property, or object around it.

Simply put, a robot that fails safely is far more valuable than one that only works under perfect conditions. Hence, make a list of as many edge cases as possible and ensure you have graceful fallback mechanisms embedded in your product to handle them. That naturally raises the question: how do you make this list of edge cases? That's where the next pointer comes in handy.

Lesson 2: Identify Early Adopters & Relentlessly Prioritise Field Testing

There is no alternative to field testing. As a robotics founder, you can spend a week running all your tests in the lab, or you can make a deal with a few early adopters and identify real customer sites that you can use for field testing.

The difference between the two is not incremental; it is exponential.

Why Field Testing Is Non-Negotiable

You may ask: why do we need field testing? Put simply, it will take ages to identify edge cases and to truly understand the impact of deploying your robot in the physical world if you rely solely on indoor, controlled testing environments or simulations. You can shorten this timeline and save vital resources by stepping out of that controlled lab and instead identifying early adopters who might be willing to both test your product and offer you a half-dozen sites to refine it. Robotics companies that have thrived over the years have followed this exact trajectory — prioritising real-world testing by identifying and establishing partnerships with early adopters rather than remaining confined to controlled lab environments.

This approach is not accidental; it is a strategic decision embedded into the operational DNA of successful robotics companies.

How Successful Robotics Companies Embedded Field Testing into Their DNA

Figure AI is a strong example of this philosophy, with BMW as its early-stage industrial partner, deploying humanoids directly on the factory floor to validate real-world use cases around automotive part manufacturing, assembly, and testing.[14] Boston Dynamics followed a similar path. Before Spot became a commercial product, it was extensively tested on real customer sites that the company gained access to through early-stage partnerships with construction firms, energy companies, and inspection teams, allowing it to refine the robot's ability to navigate and adapt to diverse terrain profiles and site constraints.[41]

Zipline offers another compelling example of this approach. Rather than confining development to controlled laboratory environments, the company partnered directly with governments and healthcare organizations in Rwanda and Ghana to address urgent medical delivery challenges, using live deployments as its primary validation and iteration platform to iterate both drone hardware and operational models.[64] Similarly, Agility Robotics began refining its humanoid Digit by partnering with enterprise warehousing and logistics operators such as Amazon, deploying selected units across their sites and conducting rigorous real-world testing focused on accelerating package handling and optimizing logistics workflows to enable commercial-scale deployments.[35]

The common thread across all these companies is clear: real environments exposed real problems — and those problems became the foundation for product maturity, reliability, and long-term customer relationships. Your goal here, naturally, is to create a win-win for both these adopters and your company. How do you do that?

How to Find the Right Early Adopters

Find customers who are desperately looking for a solution that could address even 20–30% of their problem and who are willing to give it a try. In other words, find customers for whom your robot-led service addresses an extremely painful and urgent need that no other solution available in the market can solve. These are your early adopters. Where do you find them?

Once you have clarity on the industry vertical and operational scale of customers for whom you are building this product, you would want to make a list of ten potential customers that demonstrate a clear pattern of adopting emerging technologies. These are risk-takers, to say the least, and are willing to give new technologies a try in the hope of addressing their most painful problems. For them, speed matters more than perfection. Hence, identify at least ten such organizations that would benefit from your service and that consistently show this pattern of adoption of emerging technologies.

Offer your product at this early stage to these customers, while making it clear that your goal is to help them address these problems and that you are still in the early stage of refining this product. Hence, you would want to use some of these sites as testing grounds to tune your hardware and software so you can better serve them in the long run. Needless to say, your pricing model and service offering for these customers should factor in access to real sites, i.e., training grounds, that these customers are providing, with the understanding that your product is still in its early stages. In simple words, throttle down your pricing for these customers and go above and beyond on the offering and risk-management front, without breaking your bank.

Clearly define objective metrics around what you will be testing and delivering, and within what timeline, for these early adopters. Transparency in execution and value creation for all stakeholders involved is critical to build trust here in the long run. That's a true win-win. That naturally leads to this question: what should you look for in these tests, and more importantly, how long should you conduct these?

How to Use Field Testing to Drive Product DNA

First, look for as many edge cases as possible — the impact of sunlight, red light, and white light on your sensors; the impact of floor textures; types of elevators and escalators; and your robot's ability to both identify and navigate around them, as well as integrate with them. Look at types of doors, door knobs, obstacles, tables, chairs, and the ability of your robot to interact with other humans onsite and integrate with other systems on-site. These are all both edge cases and opportunities to drive your product roadmap in the right direction.

The data and insights you collect from these field tests will help you exponentially identify an array of edge cases related to site conditions, traffic, lighting, sensors, and beyond. Now, once you have a thorough list of these, collated through extensive testing that you have conducted across half a dozen sites for each of these ten early adopters, you will need to make a strategic decision around which of these edge cases is extremely high-risk and should be mitigated through a corresponding solution that fully addresses it and therefore deserves a feature development cycle on your product roadmap, and which of these is high-risk but cannot be fully mitigated or resolved through a hardware or software-based solution.

For the latter, you will need to put guardrails in place and either cordon off areas where this edge case might show up or clearly communicate the same to the customer. So, at the end of the day, you are using this field testing to identify which edge cases should drive your product roadmap and which should be addressed through guardrail-based solutions. Examples of guardrail-based solutions include adding no-robot zones on the ROS map, saying no to sites where a large number of these edge cases exist, and so on. Your job is to stay away from scenarios that require guardrails or demand solutions that you simply cannot develop with the resources you have in place.

This ensures you are offering a product that has not only undergone mission-critical testing but is also well-defined and highly reliable. It demonstrates that you have done the deep work required to understand both the problems your product is capable of solving and the corresponding risks associated with its deployment. More importantly, it shows that you have put appropriate solutions in place to mitigate those risks, clearly communicate these limitations, and proactively steer away from sites that are not suitable for your product.

This naturally leads to the next question: what level of field testing is required, and how do you define a clear threshold for time and capital for it?

How Field Testing Becomes Your Competitive Edge

One way to think about how you want to invest your time and capital in this process of field testing with early adopters is that it's an investment to reduce future debt proactively. This means, in the first three to four years of product development, you are better off conducting as many field tests as possible across a diverse set of customer sites. You may ask, why is that the case? The reason being, at this point, your job as a robotics founder is not to blindly chase customers but to pursue those that offer you a diverse set of sites to stress test and proactively evolve your product and service.

Your goal is not necessarily to maximize revenue in the first 3–4 years. What matters more is making your product mission-critical. This means using field testing to drive both hardware and software iteration cycles, clearly identifying which problems represent painful, high-priority needs for your customers and must be solved, and which are not critical for the majority and therefore should not be prioritized. Alongside this, you must differentiate between risks that need to be proactively addressed through software development process and those that must be communicated and eliminated altogether by steering away from such sites that present edge cases your product cannot support in its current stage.

In simple terms, field testing reflects how deeply you understand your product: its capabilities, its limitations, and its ability to manage risk with precision and clarity. It also defines how effectively you translate these insights into enabling hardware and software maturity, while adding transparency and extreme clarity to your communication channels across marketing, engineering, and sales teams, ensuring everyone understands the product inside out.

You are better off conducting as many field tests as possible, especially during the first two to three iterations of your hardware cycles, even if this comes at the cost of capital and time. It is much harder to proactively identify edge cases and address those — especially if they require a hardware shift — once you have acquired customers and started scaling. Prioritise this exercise early in your journey. To summarise this lesson, product excellence in robotics ultimately boils down to how reliable your product truly is — and that leads us to the next pointer.

Lesson 3: Reliability Beats Both Autonomy and Novelty

A robot that performs reliably at 95% uptime is far more valuable than one shipped with advanced features like elevator integration but struggles to move beyond an 80% reliability threshold. You may ask: why is that the case?

Based on my personal experience, it takes a simple unplanned high-risk manoeuvre down a staircase for your robot to erode customer trust overnight and make them question your ability to understand the risks associated with your product and your capability to address them. The same applies to video feeds that your customer might be using from the camera located on your robot to monitor their site. One camera goes down for 30 minutes and that is exactly the time that matters to them — you are out. Don't let that happen. Not that customers will disengage with you overnight; there is a layer of trust that you have established between your team and your customer, but incidents that signal a lack of product reliability can be extremely dangerous for the survival of your company in the long run. Address each of these proactively through field testing and face those that show up at the last minute with urgency, ownership, and a solution-oriented mindset. That naturally raises a question: how should founders monitor and elevate product reliability beyond field testing, especially once robots are deployed in the real world?

One way to think about this is that, as a founder, you want to track both operational metrics and high-risk metrics for your fleet of robots with extreme precision and consistency. High-risk metrics include incidents where your robot was involved in scenarios that could have led to dangerous outcomes, or that directly resulted in one. These are signals you cannot afford to ignore. On a monthly and quarterly basis, you should review these metrics to ensure they are being proactively identified, monitored, and throttled down before they evolve into systemic failures.

Similarly, operational metrics include tracking uptime for your fleet of robots over extended periods across a subset of customers offering diverse site and traffic profiles. In addition to uptime, you want to track failure rates and recovery times for each core component of your robot and the fleet as a whole with extreme precision, as these collectively define the true health of your system in real-world conditions. Any hardware or software bottlenecks that degrade these metrics must be aggressively addressed, as they have a direct impact on product reliability and maturity.

At the end of the day, this exercise is not about tracking safety metrics — it is about building a deep operational understanding of how your product truly behaves in real-world conditions. That leads to our next pointer, where many founders tend to get stuck almost on a daily basis: balancing the need to elevate product reliability versus accommodating customization requests from customers.

Lesson 4: Standardisation Enables Scale, Customisation Kills It

Customisation is hidden debt. Be highly aware of it. If you are just getting started and have not yet secured multiple large-scale contracts, get comfortable saying no to unnecessary customisation requests that you may receive from your prospects. Every custom feature request will consume a disproportionate chunk of the capital you have raised and most importantly your talent horsepower. This is one of the most expensive traps in robotics and is rarely spoken about openly.

In our case, we received a large number of requests to integrate our robot with multiple systems — elevators, escalators, camera systems, barcode scanners, door systems, on-site machinery, and security monitoring systems. My approach in these moments has always been simple: you don't owe the customer an immediate answer. Your job is to carefully collate these requests, step back, and think deeply about them.

Ask yourself: if you were to develop this custom solution, how many of your existing customers would benefit from it immediately? How many would actually be willing to pay for it? If the answer is less than 80% of your core customer base, you are better off saying NO in most cases. If you don't, you will end up sacrificing your most vital resources — time, capital, and your team's energy — on custom solutions that contribute little or nothing to your financial runway. Be extremely aware of these customization traps, especially in robotics.

You should instead prioritise customers and deployment sites that offer consistent operational profiles. If it helps, think about Costco warehouses, Target stores, gas stations, oil refineries, factories, etc. Each of these sites offers a consistent operational profile, meaning the solution you develop to accommodate one Costco warehouse will, more or less, be the same for the remaining sites. The same applies to Target stores, factories, and refineries.

If you choose one highly specific vertical to target and then scale within it — subsequently adding more customers whose sites mirror the operational characteristics of that vertical — you naturally lower operational costs, minimise risk, and accelerate your ability to deliver value to customers, thereby enabling organic momentum for your company.

To summarise, the long-term success of any robotics company rests on two non-negotiables: standardised offerings and extreme product reliability.

Lesson 5: Integration Not Autonomy Defines Adoption

Now, as you approach this principle of standardisation, it goes without saying that, in the world of robotics, you won't go far if your robot does not offer APIs — which, in practical terms, boils down to your robot's ability to integrate with other systems. While these two principles of standardisation and integration might appear self-conflicting at first glance, they are not when applied sequentially and thoughtfully.

What this implies is that integration in robotics is not optional; it is a core need for most customers. Robots must be able to communicate with elevators, access systems, cameras, networks, and building protocols — quite literally. These integration capabilities often determine customer satisfaction far more than the sophisticated features that you might be shipping with your robot. Robots that do not integrate seamlessly into a client's ecosystem create friction, and friction limits adoption.

This is where discipline matters. You must do your homework. You need to identify which perception ecosystems — such as cameras, thermal detectors, and sensor suites — as well as which elevator systems, access control systems, and infrastructure stacks are most commonly deployed by vendors across your target customer sites, and then develop integration capabilities that are broadly applicable across the majority of these ideal customer profiles (ICPs).

Your goal is not to build fragmented, one-off integration capabilities, but to create structured integration pathways that scale across multiple customers — enabling both interoperability and operational consistency at the same time. You must plan integrations early in your journey as a startup founder, or you will pay for them later at a significantly higher cost because in robotics, adoption is primarily driven by how seamlessly your robot fits into the systems it must operate within.

That naturally implies the importance of paying close attention to how you design your team. Given the demands of integration, standardisation, reliability, and risk management, you need a team that is well-rounded in the truest sense — capable of operating effectively across each of these critical fronts.

And that leads us to our next suite of lessons on team design.

Team Design

Lesson 06: Hire System Thinkers, Not Specialists

This is probably the number one principle you must keep in mind as you build a robotics startup: you must build teams of system thinkers, not feature specialists. Put simply, as a robotics founder, it is always tempting to hire someone who excels at integrating robot arms or developing a strong perception system for your robot. That may seem logical in the moment, but in reality, this is one of the most common mistakes founders make when building their teams.

Why System Thinking Is Non-Negotiable in Robotics

There will indeed be times when you need people specialising in specific areas such as robot arms or battery systems. However, if you step back and study the current and future trajectory of your startup, you will realise that what you truly need in the long run is a strong team of system engineers, not specialists. You may ask why that is the case.

The reason here is simple: in robotics, product reliability and risk management are far more important than any other sophisticated feature that you may intend to offer to your client. Now, to continuously elevate product reliability and manage risks, you need folks that deeply understand all the components and sub-systems of your robot inside out: the hardware, the software, the perception system, the locomotion system, the navigation system, the motion control system, the frontend, the backend, etc. This list also includes the networking module, security module, battery system, and more. You may ask why that is the case. Why do we need folks who understand all of these sub-systems? The reason is that, in robotics, failures rarely originate in isolation.

A motor fault might be triggered by flawed control logic. A navigation error could stem from a group of sensors that lost calibration. A faulty perception output may be the result of assumptions made in the fusion layer.

The Hidden Cost of Hiring Feature Specialists

Now, if you only have specialists on your team, people who understand only their specific module, your team's ability to debug these issues slows down drastically, and so does its ability to take full ownership of each of these issues. Fragmented ownership, i.e., accountability, is a direct outcome of teams laden with specialists, and this almost always results in catastrophic outcomes. You don't want that to happen.

You need people who can take end-to-end ownership of critical failures and high-risk scenarios and resolve them with a sense of urgency. This means you need a team of system thinkers, people who can take ownership of each of these fires. That raises another question: when should you hire specialists, and how many?

While this entirely depends on the context, i.e., your current needs and the capital resources available at your disposal, a good way to think about this is to minimise the need for specialists from the get-go. How do you do that? If you noticed in the previous array of lessons on product design, I stressed the importance of product reliability and standardisation. This means even if you hire a specialist, you will need them only until you build a product that is reliable and standardised for all your customers. Post that phase, you will only need specialists on a part-time or as-needed basis. This approach will help you ensure extremely high operational resilience and capital efficiency. Some of you may ask: Does that imply depth doesn't matter in robotics?

Building Teams That Can Own Failure End-to-End

Depth is important in robotics, but cross-system breadth determines speed and survival. System thinkers understand how each critical subsystem, including perception, locomotion, navigation, motion control, battery, networking, and security modules, is developed and how these components interact to support reliable robot fleet operations. They can trace failures end to end across each of these sub-systems, anticipate second-order effects, and respond with context and urgency. These are the engineers who, at the end of the day, ensure your robots thrive in the real world. The contrast between how specialists and system thinkers operate and respond to these situations is clear in practice.

A specialist says, "This is not my module." A system thinker says, "Let me trace the root cause across the stack." The first creates delay; the second restores stability. In robotics, you need people who reason across boundaries, not just within them.

In summary, your first responsibility as a founder is to cultivate system-level thinking across your team. If you have hired specialists, make efforts to at least try to get them to mature to the next level, i.e., system thinkers. Because at the end of the day, you will rarely lose a customer or contract for not shipping a sophisticated feature in time, but you will definitely lose one if your robot suffers an outage for an extended period or runs into high-risk scenarios that lead to disastrous outcomes for the client. In these cases, you will need system thinkers and not specialists. And once you have such system thinkers in place, the next critical question becomes how quickly your organisation can respond when a robot suffers an outage or malfunctions. That leads us to our next lesson.

Lesson 07: Build Teams That Respond at the Speed of Light

Failures in robotics are highly time-sensitive and often carry disproportionate risk, which makes speed of response mission-critical. A robot stalled in an emergency corridor of a hospital, a navigation bug triggered in a warehouse robot handling 10,000 packages per day, or a malfunctioned safety sensor installed on a robot deployed on an active production line responsible for manufacturing automotive components cannot afford hours of delay. In such environments, every minute matters.

Delayed action here amplifies damage, erodes customer trust, and significantly increases operational risk. Teams that respond slowly in these cases will suffer compounded consequences. By contrast, teams trained for mission-critical response handle such failures decisively, preventing them from spiraling into crises and demonstrating the operational maturity required to thrive in high-stakes environments. So, what do these teams and their operational philosophy look like in practice?

In practice, high-response teams operate with predefined escalation protocols, clear decision ownership pathways, and rapid mobilisation capabilities. Engineers know exactly who takes control, who diagnoses the issue, and who communicates with the client the moment an incident occurs. For example, when a robot unexpectedly stops due to a sensor anomaly, such teams activate a mission-critical response chain within minutes, isolating the failed subsystems, restoring operations, and communicating transparently with both the customer and internal teams before the disruption escalates.

Ultimately, for teams driven by a mission-critical response philosophy, speed becomes an organisational reflex. Teams that internalise this reflex not only mitigate damage faster but also signal operational maturity to customers and stakeholders. And once the required response velocity is established, the next layer of safety in these operations comes from knowing when humans should override decisions made by robots. In other words, when to trigger a human-in-the-loop mechanism and intervene over the default autonomous systems deployed on your robot. That brings us to the next lesson.

Lesson 08: Human-in-the-Loop Is a Safety Strategy, not a Choice

Full autonomy remains imperfect, and assuming otherwise introduces unnecessary risk. Human-in-the-loop mechanisms are therefore not optional safeguards; they are foundational to safe and responsible robot-led deployment. Teams must clearly define when human intervention occurs, how it is triggered, and which individual or sub-teams are ultimately responsible for accountability and ownership in such cases. No robot should operate without clearly established ownership structures and escalation protocols guiding these critical moments of control transfer.

This need becomes even more evident and equally critical when you consider the environments in which robots operate. From warehouses and offices to oil refineries, mines, and tunnels, robots today function across a wide range of indoor and outdoor settings. As a result, they will inevitably encounter conditions they may not have been explicitly trained to handle. From unexpected human behaviour and terrain variations to sensor degradation and system anomalies, the edge cases in these environments are virtually infinite.

Without clearly defined protocols to transfer control from robot to human, failures in such scenarios will not only escalate but can, in many cases, lead to disastrous consequences. Robotics companies that have successfully navigated these challenges have therefore proactively embedded human-in-the-loop systems with clearly defined escalation points and control handover mechanisms. From Waymo and Amazon to Boston Dynamics and NASA's JPL, this approach has become a core safety and control strategy across the robotics industry. As a robotics founder, once these safety foundations are firmly established in your operations stack, the next challenge really boils down to how humans emotionally and psychologically accept the presence of your robots in their environment.

Lesson 09: Psychological Adoption Is More Important Than Technical Delivery

The lesson is simple yet often underestimated: no matter how advanced your robot is, it will fail if people do not feel comfortable using it. Robots operate in social environments, and adoption is shaped as much by technical excellence as by emotional comfort and psychological acceptance.

This matters because fear, mistrust, and confusion can quietly sabotage even the most capable systems. A robot that moves unpredictably, appears intrusive, or feels difficult to understand creates emotional resistance. Users hesitate, avoid interaction, or bypass the system altogether. Over time, this erodes confidence, reduces utilisation, and limits long-term value — regardless of how well the underlying technology performs.

Consider a robot deployed in a hospital, factory, or retail environment. Even if its navigation, perception, and spatial manipulation modules are designed flawlessly, staff may feel uneasy if they do not understand its behaviour or feel unsure about how it will respond in critical moments. They might tolerate the robot but will never truly accept it. Tolerance here never translates into long-term adoption. So how do you address this challenge?

A Two-Pronged Strategy to Drive Psychological Adoption

The way to mitigate this challenge at a psychological level is by adopting a two-pronged strategy. The first part of this strategy is the design and feel of the robot itself, and the second is the way you introduce it to your customers and onboard them. Speaking of design, it should look, feel, and perform in a manner that is intuitive and human-centric. The choice of color, the shape of your robot, the warmth it conveys, the lights installed on it to communicate its intent, the sound systems, and the ambient sounds that ensure others are aware of its presence — all of this matters. Unless you reach this point, long-term and wide-scale adoption will remain a major risk for your company. That is only part one.

The other half of this equation is onboarding. How you introduce your robot to your customers matters. While there are many ways to do this, one approach that has worked well for us is doing this in a phased manner. First, you build alignment with your decision-makers and make them feel comfortable about adopting your robot. You do this by demonstrating its performance in real time and proactively addressing concerns around safety and risk. Then you do the same with the immediate staff who will be working with the robot, and finally with the organisation as a whole. These three are your core engagement points to ensure your robot is adopted at a psychological level — not just by decision-makers on the customer side, but by the organisation as a whole.

In summary, robots succeed when the people adopting them feel safe, informed, and in control. Technical capability enables operation, but psychological acceptance enables adoption. And once adoption is secured, the next critical foundation lies in ensuring you have the right team in place at all times.

Lesson 10: Team Durability Is Your Number One Priority

This is probably the hardest thing to execute in reality, and one that carries profound implications for your startup. The lesson here is really about your ability to bind your core team members and retain them in the long run. The only organic way to achieve this is to create a win-win for both you and these core individuals. In theory, you want to do this with everyone in your company, but that becomes extremely difficult as the organisation scales. Hence, a practical starting point is to identify those core team members who are absolutely critical to your company's long-term success and create win-win pathways with these folks at a deeper level. The question naturally is: how do you do that?

A Three-Phase Framework for Building Your Core Team

You will need a three-phase strategy here. In phase one, you must identify these individuals.

Phase I: Identifying Potential

You cannot simply identify them based on the seniority of their role or their current performance. You need to observe and evaluate them over a period of at least 12 months to truly understand their performance, work ethic, and their philosophy of life. Once you have closely followed them for a year, you will be in a far better position to filter them. Make no mistake, some of these individuals could be working at the periphery of your company — technicians, part-time staff, associates — and some might even be in their 60s. Do not let age, seniority, or role guide this filtering process.

Phase II: Evaluating for Consistency, Ownership, and Integrity

Assess them solely based on their performance and their ability to take ownership more for failures and less for successes. One point of clarity here is that these individuals carry a core trait: they are highly consistent across all fronts — work, life, and health. Beyond this core trait, there is a master trait they will exhibit almost without exception: the ability to take ownership of failures. They will be the first to raise their hand to handle fires, and many will take the initiative to step in even without being asked, because they genuinely care about what they are building and will go above and beyond to protect it. That level of integrity and consistency is what makes them stand out naturally.

The next phase is to offer them a level of responsibility that you, as a founder, can comfortably entrust them with — typically at least 2x more accountability than their current role — even if this introduces a degree of risk due to the decisions they may now have to make. Observe them carefully over the next six months and mentor them as needed. Based on their overall performance and ability to make decisions, you will then be in a position to move those who pass this stage to the final phase.

Phase III: Scaling Responsibility and Securing Long-Term Alignment

The last phase is to have an open and honest conversation with these individuals about their long-term career trajectory and determine whether a true win-win relationship can be established. There is really no alternative to this phase other than having a serious and open one-on-one conversation with each of these individuals. Thus, your end goal is to ensure you have a strong core team built through a deliberate process of identification, evaluation, and long-term alignment. This is the only realistic way to enable sustained operational resilience and true team durability. The rest — centralized documentation, training sessions, and workflows — are largely cosmetic without this foundation. When this foundation is in place, your organisation gains the operational endurance required to scale, adapt, and succeed in this arena in its truest sense.

Now that we have covered the lessons on building high-performance teams and products to enable operational endurance for your robotics startup, the next question that naturally follows is: how do you communicate the exact value that your product or services offers to your customers and build trust? In simple terms, what are the core lessons you must keep in mind as you develop your sales and marketing strategy? We will cover these in the next chapter.

Chapter 23: Sales and Marketing

Hard Lessons on Delivering Value and Building Customer Trust

As a founder of a robotics startup, this is probably one of the topmost operational areas that keeps you awake at night, and rightfully so. It has a direct impact on your ability to build trust with your customers and investors. To do so in a sustained manner over a long period of time is extremely hard, but not impossible.

There are a gazillion paths you can take to build your sales and marketing strategies, and each is laddered with crevices so deep that they will cost you a ton of your time, resources, and talent horsepower. From defining your client acquisition strategy to being selective about trade shows and conferences, and navigating the security maze when working with enterprise prospects, these are some of the most risk-prone areas that you will need to navigate almost on a daily basis. If you do not take the time to study and develop a surgical approach to address each of these proactively, they can bring your startup to a standstill in no time.

The question is: how do you do that? In this chapter, we address this question through a series of hard-won lessons and the corresponding practical frameworks you will need to tackle each of these challenges, one by one.

Lesson 01: Sell the Problem You Solve, Not the Robot

The first lesson on the sales and marketing front that I would like to emphasize is that even though your solution is represented in the physical form of a robot, the robot itself is not the solution you are selling. Your customers do not care whether you solve their problem using a robot or a camera; they care about how effectively you are able to address their pain points. Your robot is only the delivery mechanism for your solution, not the solution itself. This is the number one mistake many robotics founders and early-stage teams make when speaking with prospects and customers at marketing events and even during demos.

As a founder, it is your responsibility to ensure that your company's messaging strategy is anchored in the customer's true pain points and clearly demonstrates how your solution directly addresses them. If you fail to do this and instead focus on the robot or its features, you are setting yourself up for failure. No one, literally no one, wants to buy a robot — but they are willing to buy one if it addresses their core pain point 10x better than any other solution available in the market. This is not theoretical advice; I have experienced this firsthand while conducting demos for companies such as Apple, Toyota, Western Digital, and more. Be aware of this lesson.

Define the Outcome You Deliver — With Precision

Define precisely what you are offering: reduced downtime for a production line, faster throughput for warehouse operations, improved safety in oil refineries and tunnels, or cost efficiency for automotive manufacturing plants. These qualitative outcomes must be supported by quantitative evidence generated during the early-stage adoption phase of your product.

In simple terms, clearly define what percentage improvement an automotive customer will see in production throughput if they adopt your robot compared to their existing solution. The same applies to warehouse operators. You must provide a quantified metric that justifies your offering and support it with evidence-driven case studies from clients who have achieved those results and are willing to speak with your prospects.

The robot is only the tool that enables this outcome.

Once you have precisely defined your offering with this metric, the next critical step in driving customer adoption is establishing trust. How do you do that? This is where the second lesson comes into play.

Lesson 02: Let Your Product Become Your Strongest Marketing Channel

Trust in robotics ultimately comes down to product reliability. A robot that performs with 99.99% reliability year after year will always be preferred over one that operates at 95% reliability, even if it offers superior autonomy and integration capabilities. The reason is simple: customers are not just buying your solution; they are buying peace of mind. They want confidence that your system consistently addresses their pain points without introducing new risks or uncertainties. As a result, most will prioritize reliability over advanced capabilities.

How Mission-Critical Product Reliability Turns Customers into Advocates

If you indeed provide the most reliable solution available in the market to address their pain point, not only will these customers prioritize and adopt your robot-led service, but they will also become your proof points and a force multiplier for your sales and marketing engine. In simple words, they will help you eradicate the friction that frequently catapults these sales conversations into a death valley of doubt and compliance, and help shorten decision cycles. You may ask: how?

While I can't speak on behalf of other companies, in our case, we had a select group of customers who were always willing to speak with our prospects and openly communicate what they truly valued about our robot-led safety and security service. In simple terms, they became our force multiplier on this front. Hence, as a robotics founder, your focus should be on product reliability, because nothing markets better than a robot-led system built with mission-critical reliability.

Now, some of you may ask: beyond focusing on product reliability, what about traditional marketing channels such as conferences and trade shows? How should robotics founders approach these platforms to drive brand awareness and product adoption?

Lesson 03: Be Selective with Conferences and Industry Events

Truth be told, most conferences are nothing more than a distraction for any startup. The signal-to-noise ratio in this channel is extremely low. Most are not worth your time. A few, however, can open doors to strategic buyers or early adopters. Before you invest your hard-earned capital in securing a booth at these conferences and trade shows, it is critical to develop a framework to identify which of these events are truly worth your time. How should you do that?

How to Evaluate Conferences and Trade Shows as a Robotics Founder

On this front, my personal experience has been that history is the best teacher. Look at the performance of each of these conferences over the last three to five years. Trace back the robotic or emerging technology companies that launched their products there, and then reverse-engineer how many of them actually announced partnerships or derived immediate business value from these channels. You may ask: what approach should a founder employ to filter the conferences that truly provide value?

A practical way to establish this link between conferences and the corresponding true value they provide to startups is to look for evidence. Press releases, post-event announcements, webinars, or presentations delivered by startups during or immediately after these conferences often reveal what led to tangible partnerships or momentum. If you do not have time to conduct this exercise yourself as a founder, assign a marketing associate to focus on this research. Doing this once and maintaining it as a reference framework is extremely valuable. It will save you significant time, energy, and operational headaches over the years. Put simply, you are far better off doing this exercise than betting $20K to secure a conference booth and another $50K on deploying your team of world-class engineers at the event for four days without calibrating what the returns will be.[65] It is a no-brainer to do this homework rather than gambling $70K of your operational capital on each of these events.

In our case, we attended the ISC West conference and quite literally shook hands with a Fortune 100 e-commerce and cloud provider based in Seattle. Their team visited our booth, spent nearly 30 minutes evaluating our solution, and handed us their cards to test our system on their campuses. That is the kind of outcome you are looking for.

As a robotics startup founder, your goal is not necessarily to establish ten partnerships at each of these events, but to use those three to four days surgically to secure one or two partnerships that are truly worth gold in the long run. Do your homework, apply the approach I have shared here, and if you cannot, at least take the time to speak with other founders who have attended these events and successfully secured partnerships. If you follow these steps, the outcomes will follow.

This brings us to the next lesson on building your client pipeline.

Lesson 04: Design a Client Pipeline That Balances Speed, Risk, and Scale

How should a robotics founder define a client pipeline? It ultimately depends on your end goal and the operational stage you are currently in. If you are at an extremely early stage, your priority should be to secure early adopters and drive field-testing efforts, as discussed in the product excellence session. If you are past that phase and have already secured a few clients, your focus should shift toward establishing a balanced mix of Fortune 500 companies (i.e., large enterprises) and mid-sized organizations.

Some of you may ask: why not just work with mid-sized organizations, especially since their sales cycle is relatively shorter and better aligned with the resource constraints of a startup? Or, on the other hand, why not work only with enterprise clients? Why do we need both?

Why You Can't Rely Solely on Mid-Sized or Enterprise Clients

Because you want to strategically balance operational speed, risk, and scale. It is important to understand that enterprise clients bring credibility and long-term stability but move at a snail's pace. In most cases, the first six months are spent evaluating your product, the next six conducting pilot programs, and the following six months executing the first production-grade deployment. You are effectively looking at an 18–24-month runway before you see tangible revenue from enterprise clients.

Mid-sized companies, on the other hand, move at the speed of light but carry higher financial and operational risk. From constrained financial runways to acquisition by larger organizations, many mid-sized companies operate at the edge of speed and instability. Hence, an optimal client pipeline blends both segments — allowing you to generate early-stage revenue velocity from mid-sized clients while simultaneously building long-term strategic momentum with enterprise accounts. This balance ensures steady cash flow, minimizes concentration risk, and maximizes your operational endurance.

As you build this client pipeline, you must remain conscious of the distinct risk profiles associated with each client type. This brings us to the next set of lessons on evaluating and managing the risks involved in selecting clients and prospects.

Lesson 05: Security Is Non-Negotiable for Enterprise Deals

The number one risk you will encounter when dealing with a Fortune 500 prospect or client is the chain of concerns they will raise around the safety and security profile of your product from day one — and rightfully so. For these enterprise companies, security is not a feature; it is a gatekeeper. If you fail their security and compliance evaluations, the deal ends before it even begins. That naturally raises a question: why is that the case? Why are security and compliance such deal-breakers for these clients?

The reason here is that these clients operate at a global scale in most cases, meaning they are not going to deploy your product at just a dozen sites. It is highly likely that if you are engaged with them, they are evaluating your product for deployment at a global scale. They want to ensure that if they sign a contract with you, they will be able to deploy your product across all their sites. In simple words, your product must be capable of addressing their underlying pain point in a consistent and reliable manner across their global footprint.

Hence, naturally, the biggest risk they need to evaluate your product or service against is safety and security. Is it safe, and more importantly, is it secure? Your architecture must therefore reflect enterprise-grade safety and security standards from day one. As a founder, security must be embedded into your product design and development philosophy from day one. That naturally raises an important question: how do you do that?

Design Your Architecture for Enterprise-Grade Security

One approach here is to evaluate your product's safety and security profile from every possible lens: hardware, software, supply chain, battery systems, and more. If there are areas that are not yet fully secured, you must clearly communicate what guardrails are currently in place and what your long-term plan is to address those gaps. Do this process in an iterative manner as you conduct your field testing, as discussed in the section on product excellence. Make no mistake — this iterative process of field testing, safety and security evaluation, and documentation will take time and resources, but it will pay off in the long run.

Communicate Product Risks & Guardrails with Absolute Clarity

Once you have done this exercise over a period of 6–8 months during the early stage of your journey, you will have a clear message justified with evidence around which elements of your product are safe and secure, how you arrived at those conclusions, and what guardrails are in place for those that are not yet fully secured. This will provide clarity and transparency to both you and your customers around safety and security. If it helps, I have already covered some of the lessons on how a Fortune 500 pharma client evaluated our robot-led safety and security service end-to-end — from software and SOC II to broader operational safeguards — in part II of this book. You may want to give it a read, if you haven't already.

While safety and security posture of your product remain some of the most critical requirements and therefore the primary risks in enterprise deals, operational longevity becomes the core risk when working with mid-sized clients. That brings us to the next lesson.

Lesson 06: Understand the Risk Profile of Mid-Sized Clients

Mid-sized companies offer an accelerated revenue trajectory and relatively shorter sales cycles compared to enterprise clients, but they come with extreme volatility. What does this volatility look like in the real world?

Why You're Always on the Chopping Block

What you will see is that some of these companies eventually get acquired by larger organizations, while others may run out of funding, inadvertently putting you and your company on the budgetary chopping block. In our case, one of our mid-sized clients was acquired, and we were fortunate that the acquiring company chose to renew our contract. In another case, we were not that lucky, since the acquiring company decided to stick with their existing vendor to enable safety and security services over our robot-led service.

These are two important lessons to remember here, since the biggest risk that comes with mid-sized clients is that they offer speed, they scale fast, and you scale with them — but when times like these come, whether it is acquisition or equivalent organizational shifts, you will undoubtedly be on the chopping block just like other vendors. If you recognize this risk early and address it proactively by structuring your contracts, revenue models, and overall client portfolio strategically, you significantly strengthen your ability to withstand these financial shocks and maintain long-term operational stability.

Now, beyond understanding the risk profile of your clients, irrespective of whether you are working with mid-sized companies or Fortune 500 enterprises, you must also evaluate the risk profiles of your vendors. Your supply chain must be equally risk-proof. You may ask — why is that the case?

Lesson 07: Your Supply Chain Is a Risk Vector — Ignore It and Lose the Deal

As a robotics founder, you are cognitively aware that you are not running just one company — you are effectively running a network of companies. You are procuring sensors, cameras, batteries, networking modules, GPUs, molds, and then developing a software system that binds all of them together to create value for your customers.

Every Vendor Introduces a New Risk Variable

The challenge is that this very process of integrating multiple components results in establishing partnerships with a series of vendors — one providing cameras, another LiDAR, another lithium iron phosphate batteries, and yet another cloud and database systems. Each of these vendors introduces a new risk variable into your equation. This is not just a risk to your product or service, but a direct risk to your client relationships and enterprise deals. You may ask: why is that the case?

Why Enterprise Clients Scrutinize Your Entire Supply Chain

If you are working with enterprise clients, they will surgically evaluate both your product and your supply chain for safety and security risks. Following this trend, many mid-sized companies have also begun incorporating similar evaluations into their sales cycles. This means you must assess your vendors from the very beginning and establish partnerships only with those that provide both clear value and peace of mind. Those that you can depend on as much as you do on your core team members. That should be the level of evaluation and hence the trust in this case.

Speaking of trust, security is one of the key deal-breakers in the robotics space, and one of the most common misunderstandings is the belief that you need a dedicated security team to secure industry-grade certifications such as SOC II and GDPR. This is not necessarily true. Irrespective of the size of your team or your resources, you can, in most cases, complete these certification processes with integrity and within reasonable timelines if you have the right approach in place. The next lesson elaborates on this topic.

Lesson 08: Navigate SOC II & GDPR with Lean Teams

SOC II and GDPR compliance are indeed often deal-makers or deal-breakers. These certifications strongly influence trust, especially in regulated industries such as healthcare, automotive, banking, and warehousing. While the need to secure these certifications is clear — to demonstrate the level of safety and security protocols built into your product and to provide customers peace of mind — many founders struggle to understand how to approach these certifications and, as a result, steer away from them and the corresponding enterprise deals. This is a big mistake.

As discussed in the earlier lessons, you will inevitably have to deal with these certification requirements sooner or later, since enterprise clients will become part of your client portfolio. So how do you address this requirement from the very beginning? Do you need a massive team? Definitely not. There are multiple ways to approach this challenge, and in our case, we partnered with a third-party organization that specialized in helping startups complete their security certifications with an extremely high degree of time and resource efficiency. They guided us through this process and helped us reach the finish line without draining our talent or capital reserves. You will find more context and details on this approach in Part II.

This later became a strategic advantage as we started engaging with enterprise clients, because compliance was consistently among the top three questions we received. Having these certifications in place from the very beginning allowed us to address those concerns transparently and with extreme clarity. This approach enabled us to build trust with enterprise clients from the outset, shorten sales cycles, and accelerate the velocity of every deal that followed this certification process. Thus, it created exponential value for both our clients and our company.

While security and compliance certifications may help you secure enterprise deals, sustained adoption post-sale truly depends on how effectively your team educates and enables customers. This leads us to the next point.

Lesson 09: Educate Your Customers Relentlessly — or Lose Them

A robot-led service is almost never a plug-and-play operation for most clients. While software systems can often be sold with structured online documentation, in the robotics arena, documentation alone is never enough — and can even cost you a client deal. You may ask, why is this the case? The reason is simple: when people see a robot, there are instantly ten questions running through their mind. These concerns revolve around sensors, hardware, safety, reliability, and performance. Most are also wondering what the robot is capturing, what is being recorded, for how long, and how it behaves. In simple terms, you are placing an autonomous mobile machine into their environment. It is equivalent to deploying a car inside someone's office space. No amount of documentation alone can extinguish the fires of doubt and perceived risk you will face during both pre-sales and post-sales periods.

So, what do you do in this case?

As soon as you secure a deal — whether with enterprise clients or mid-sized organizations — make it a priority to conduct a three-phased product introduction and adoption program. In simple words, your customer onboarding protocol must be surgically crafted with visuals and real-world evidence showing what your robot does, how it behaves, what it is not capable of doing, and, most importantly, what safeguards and systems are in place to ensure it causes no harm. The more precise and disciplined you are in these customer education sessions, the deeper the confidence you instill in them and their organizations as a whole.

This exercise, when conducted repeatedly, for example on a quarterly basis at the beginning, then at the six-month mark, and again after a year, will ensure higher client retention rates.

Beyond building a great product and assembling a strong team, this is likely your next most important responsibility as a founder. Because in robotics, while securing a deal is hard, retaining it is even harder. Most deals are either truncated before the contract period ends or never renewed because founders and their teams fail to establish deep trust around the safety and security profile of their product in the minds of both decision-makers and the organization at large.

That leads us to our next point on client retention.

Lesson 10: In Robotics, Retention Is Exponentially Harder Than Sales

Closing a deal is only the beginning of the sales marathon you will run in robotics. The hardest part is truly retaining these clients in the long run. You may ask, why is that the case? There are numerous reasons why a client may truncate a deal or decline renewal.

In the robotics space, the most common reasons clients cancel or do not renew are the following. First, you and your product failed to deliver the value your client expected — and worse, you did not identify or resolve this issue while the contract was still active. Second, while decision-makers might have been satisfied with your product, the organization where your robot was deployed remained uncertain about its safety and risk profile, raising concerns at scale. Third, a high-risk incident involving the robot may have led to immediate loss of trust and cancellation of the contract. Last but not least, your client may have been acquired by another firm, and the acquiring company chose to proceed with a different vendor due to pre-existing partnerships.

I have seen many deals collapse due to these exact reasons, based on my experience working with clients across the robotics spectrum — from pharma and warehousing to automotive manufacturing and cloud technology companies. Whatever the cause of a lost deal, each of these pain points can be proactively addressed if you reverse-engineer the core reasons behind them and tailor your sales, marketing, engineering, and operational protocols accordingly. That naturally leads to the next question: how do you do that? How do you identify these pain points proactively before they cost you a contract?

One approach here is to establish a strong two-way communication channel that you can activate at least twice a month with every client that has signed a contract with you. You must ask difficult but necessary questions during these sessions: how satisfied are they truly with your product, and what are the top three things they would like changed to improve value and peace of mind? If a client has no pain points to cite, or worse, is shying away from providing feedback, it is your responsibility to ask the right questions and get that answer. You need to surgically identify points of concern and dissatisfaction early and address them proactively if you are serious about retention.

Client retention must be treated with the same level of priority as securing new clients in the first place. That brings us to our next lesson.

Lesson 11: A Concentrated Client Base Is a Single Point of Failure

As a robotics founder, based on the lessons covered in the earlier chapters of this book and throughout this section, one truth becomes clear: diversification is not merely a growth tactic, it is a survival strategy. You must intentionally diversify your client base across both industry verticals and company scale.

Relying heavily on a single industry or a narrow client segment exposes your startup to concentrated risk. Economic downturns, sector-specific slowdowns, or regulatory shifts can quickly destabilize your revenue pipeline. A resilient robotics company distributes its presence across multiple verticals and client types, allowing it to absorb these shocks without crippling its core operations.

In our case, we worked with clients across the technology sector, warehousing, and automotive manufacturing industry. Beyond industry diversity, our client base was also distributed by company size — some in the mid-sized category and others at the enterprise level. This level of diversification proved invaluable when several of our mid-sized clients were acquired and our contracts were prematurely truncated, while some of our enterprise clients were impacted by industry-vertical-specific volatility and opted for aggressive budget cuts, placing us on the chopping block.

This diversification strategy not only protects your company against market-related volatility but also enables long-term operational and financial endurance.

Now that we have covered the lessons around sales and marketing, focusing on what founders should proactively keep an eye on as they evaluate prospects and build a balanced client portfolio, it goes without saying that, in the long run, the success of your company will predominantly depend on how effectively you retain those clients and, more importantly, how consistently you deliver sustained value to each one of them as you continue to add new clients to this ecosystem.

In simple words, your operations team will play a critical role as you scale your company, and they will become the heartbeat of your revenue engine. The question is: how do you build a strong operations engine to steer and scale your startup? We will address that question in the next chapter.

Chapter 24: Operations

Hard Lessons on Building a Robust Execution Engine

Daily fires—that's probably the simplest way to describe what operations looks like at a robotics startup, and that's not an exaggeration. Unless you are already operating at enterprise scale, once you onboard your first few customers, you quickly find yourself dealing with fires on an almost daily basis. You might ask: what kind of fires? There are many, to say the least. It can start with something as routine as a software release pushed by an engineer to your fleet of robots. One of those robots happens to be a demo bot, running live in front of a customer. The release goes wrong. What follows is not just a failed demo, but a fleet-wide outage triggered by that same release.

Or consider a different scenario: an unintended high-speed run caused by a malfunctioning sensor located on a robot navigating a ramp at a customer site. Soon after, you receive a call because your robot's localization module has lost calibration, and the robot has started drifting into places and corners it should never enter. These are the first-order effects—the incidents themselves. The second-order effects are equally dangerous: not having enough people, time, or resources to diagnose and resolve each issue in time. The third-order effect is the hardest of all—building the systems needed to monitor each of these failures and proactively add guardrails to prevent them from the get-go.

Operations, in reality, is about managing all three orders of fires simultaneously. The question, then, is simple: how do you do that? In this chapter, we answer that question through a set of pointers and practical frameworks you can apply on your journey from day one.

Lesson 01: Build a Rapid-Response Flying Squad for High-Impact Demos

In the early stages of your journey as a founder of a robotics startup, your ability to demonstrate value in real environments, i.e., directly at the site of prospective early adopters, determines how effectively you are able to convert them into paying customers. This is much easier said than done in the field. You may ask, why is that the case?

After having conducted more than 100+ robot-led demos for customers around the world, one thing I have realized is that each of these demos consumes a significant amount of your vital resources — time, capital, and talent horsepower. Hence, you need to develop a surgical protocol to execute each of these without exhausting your resources. That naturally raises the question: how do you do that?

One approach that worked for us during the first few years of our journey in this field revolved around the concept of developing a highly responsive demo team, i.e., a "flying squad" capable of executing on-site demos with precision, speed, and extreme consistency. That naturally raises another question: how do you develop such a team? Who should be on this team and why?

Based on numerous iterations that we did on this front during the first few years, we ultimately realized that the most effective team structure is one that is extremely lean and focused: a dedicated demo engineer supported by region-specific operations managers who coordinate logistics and client relationships. In simple words, you will need to dedicate one all-rounder engineer who can conduct all of these demos and can be paired with operations managers allocated across different geographies based on the countries you are targeting.

In our case, we had about half a dozen operations managers responsible for managing global operations and one dedicated engineer to conduct these demos and deployments. After executing an array of demos, we realized that this model maximizes your ability to say "yes" to as many relevant prospects as possible, i.e., your demo capacity, without overloading engineering resources. Without this structure in place, or with a structure that leans too heavily on engineering resources, you will create bottlenecks for your company, either on the sales front or on the product excellence front. So, you must adopt a surgical approach to determine the number and type of resources, especially during the early stage, you are willing to allocate to this exercise, as it is vital for your survival and directly impacts your revenue cycles.

That naturally leads us to our next lesson on dedicating hardware and software resources for your sales meetings and product demos.

Lesson 02: Maintain Dedicated Demo Environments — or Pay the Price

One of the most overlooked operational mistakes founders make in early-stage robotics startups is that they inadvertently allow demo-related operations to interfere with their live production environments. In simple words, it is highly tempting to maintain a single fleet of robots and one software environment for both production i.e., post-sales deployments and pre-sale demos. That is a big mistake. The reason is simple: if you do not isolate them, you end up introducing far more risk into your operations protocol than is necessary. You may ask, what kind of risk?

Why Demo and Production Environments Must Always Be Separate

Well, you will inevitably face a range of operational risks and failures if you do not separate your demo and production environments, both at the hardware level i.e., a dedicated robot for demos and at the software level i.e. a dedicated setup of your web application for customer demos. We experienced this firsthand when several of our early-stage demos were negatively impacted due to a last-minute software release that was meant for the production fleet, i.e., robots deployed at live customer sites but was unintentionally pushed to the demo robots as well. This caused the robot to stop working in the middle of a live customer demo, and that is almost always a red flag. No matter the reason, if your robot stops functioning mid-demo, you are already on the chopping block in the eyes of the prospect. So be extremely careful here. This is just one risk; there are many more that follow this same pattern.

Without this deliberate separation of hardware and software environments for pre-sales and post-sales operations, failures in demos can cascade into live production environments—and vice versa—exposing your organization to unnecessary operational risk and lasting reputational damage. That brings us to our next lesson on product demos.

Lesson 03: Offer Both Remote and In-Person Demos — or Lose Deals

As a robotics founder, it goes without saying that you should prioritize in-person demos wherever possible. Nothing builds trust in your product faster than seeing the robot navigate an office space or warehouse in person along with the corresponding insights it offers to the customers. However, there will be cases where in-person demos are not feasible, and you will need to conduct them remotely.

You may ask: what kind of scenarios lead to this choice?

Why Remote Demos Matter

Once you are in the field, you'll find that this can happen for many reasons. Sometimes it's a resource constraint. You simply cannot have multiple teams flying out in parallel for in-person demos. Other times, it is the prospect who is not yet ready to host you on-site or has not developed enough trust in you and your product to offer an in-person meeting with their team. Either way, you must be prepared to adapt. As a founder, you must be agile enough to accommodate these scenarios from day one.

That naturally raises the question: how do you address this challenge?

What should you do in these cases?

How to Conduct Remote Demos

You must maintain a dedicated remote demo setup for scenarios where the prospect is still exploring your solution and has not yet opened the door for an on-site demonstration. In such cases, your goal should be to demonstrate how your robot solves core pain points for your existing customers in real time by setting up a controlled demo environment within your lab or office space. You should be able to live stream both the robot's movements and the corresponding insights it delivers in real time, allowing your prospects to observe these two feeds in parallel on their computer.

Your prospect should be able to see, side by side, how your robot moves, how it detects obstacles, how it handles edge cases, and the true value it creates for them through the insights delivered via your dashboard remotely.

In addition to this real-time, remote-first demonstration, you must also share detailed case studies that clearly demonstrate how you are creating value for your existing clients. These efforts will help your prospect develop deep trust in your service from day one.

In simple terms, your remote demo must help prospects clearly visualize what it would look like to deploy your robot and software in their environment and how it would alleviate their most critical pain points while providing them with peace of mind.

Mitigating Risks with a Dual-Mode Demo Strategy

This dual-mode approach for your demos that blends in-person and remote capabilities will help you expand your reach, accommodate more prospects, accelerate decision cycles, and reduce friction in your sales operations. If your demo strategy is limited to a single mode, you will inevitably lose some of these time-sensitive deals. Be aware of this trap.

Now that we have covered several lessons on the pre-sales front, it is important to remember that your operations team will play an equally critical role on the post-sales front. That means, once a deal is signed, ensuring consistent product performance and reliability at the customer's site becomes a top priority for this team. That is where the next lesson becomes especially important.

Lesson 04: Fleet KPIs Must Guide Your Sales, Product, and Risk Strategy

Once your robots enter the field, performance data becomes your most valuable operational asset. You may ask: why is it so important? More importantly, how does it matter in the long run?

Because this data captured from the field directly informs your sales strategy, your product development roadmap, and significantly improves your overall risk profile — three pillars that are equally critical for retaining clients over the long term. We will now cover each of these one by one.

Qualify Better Sales Deals

To begin with, the data you collect from the field helps you identify the customer profiles and site types where your robot performs best. These are environments where it thrives, creates exponential value for your customers, and encounters minimal high-risk scenarios. Armed with these insights, your sales team can qualify prospects with a surgical precision and confidently say no to deals that are not a good fit for you given your product's current performance and reliability profile. This, in turn, helps minimize churn, compress sales cycle, and accelerate deal velocity. Put simply, it makes your sales team ultra-efficient.

Elevate Product Reliability

Beyond making your sales team ultra-efficient, it also makes your product development team equally focused and efficient. You may ask how. This field data helps you identify your absolute highest-priority areas for addressing risk in your product, thereby enabling you to drive your product development roadmap in a more methodical manner.

For example, if you are able to trace a failure in your perception stack all the way down to a camera that, let's say, doesn't perform well in the presence of bright lights, that anomalous behavior will show up as a recurring pattern across your fleet, and this fleet data will help you identify the root cause. Thus, this data will guide your choice of components—sensors, cameras, batteries, motors, displays, onboard computers, database systems, and cloud infrastructure—as well as your vendor evaluation process for sourcing these critical elements. Thus, it will help you elevate the safety and reliability profile of your robot. This process is extremely important in the long run because a robot that is safe, secure, and reliable will always be preferred by clients over one that is loaded with sophisticated features but delivers subpar reliability.

Simply put, this field data will help you make informed decisions for your product. It will allow you to prioritize product features that are truly critical to both you and your clients.

Strengthen Market Positioning and Enable Long-Term Stability

Last but not least, this data will help you understand the long-term risk profile of your product across industries. It enables you to double down on industry verticals where your product performance is strongest and disengage from those that consistently signal high risk for you and your clients. In our case, this field data helped us understand that our security robot, in its current state at the time, was not truly ready to patrol automotive manufacturing plants but was extremely well suited to patrol indoor office spaces for technology clients. This level of clarity is exactly what you are looking for.

This is extremely important because, as a founder, you will now be able to drive all your teams — sales, engineering, and operations — in one direction using this data, ensuring that your company remains laser-focused on this vertical. This in turn will help you reduce churn, stabilize your revenue channels, and enable long-term operational resilience for your company.

Now, given these three pillars, the next question is: how do you monitor this data and the corresponding metrics effectively?

Which Fleet Metrics Matter Most?

While the exact metrics may vary based on your use case and industry vertical, some metrics are universally critical in robotics operations. In terms of what you should track, focus on the metrics that directly impact safety, uptime, and operational endurance. These include robot uptime versus downtime, Mean Time Between Failures (MTBF), Mean Time to Resolution (MTTR), battery health and degradation patterns, sensor failure rates, navigation errors, network stability profile, autonomous task success rates, and the frequency of emergency stops or safety incidents. These metrics give you a real-time pulse of your system's health.

You must first define a clear set of KPIs for your fleet from the very beginning. The way this should be done is by following a staged approach.

Defining Fleet KPIs: A Three-Stage Approach

Stage I is really all about getting into the weeds, and this means each team will have its own set of metrics that allow it to dive deep into its area. Think about frontend metrics, backend metrics, robot-related metrics, and so on. These are your Stage I, or deep-dive metrics.

Stage II consists of more executive-level metrics. These are the overarching metrics that should matter to you as a founder and to your executive leadership. They should help you understand the overall reliability and value-creation profile of your system without burdening you with all the details. For Stage II, we are looking for no more than 5–6 metrics at the fleet level. These will include overall uptime across the fleet per month, high-risk incidents per month, task-completion rate across the fleet, and similar top-level indicators.

These KPIs will guide the types of sales deals you pursue based on your product's current reliability and performance profile, while also enabling your operations and engineering teams to elevate product reliability. That naturally raises another question: what tools should founders use to track these metrics?

What Tools Should You Use to Track These KPIs?

To do this well, you need a well-defined system that allows you to see how your fleet is performing in real time. This is where a structured dashboard becomes critical. Depending on your stack and scale, tools like Datadog, Grafana, Prometheus, New Relic, Splunk, AWS CloudWatch, or Azure Monitor can provide this capability. In our case, we used a combination of AWS tools to monitor our cloud components and Datadog to monitor robot components, battery failure rates, and related performance metrics. That raises another question: what is it that you should optimize for as you build this dashboard?

The objective behind putting this dashboard together should be to help you and your team visualize fleet data in a way that allows you to spot anomalous behavioral patterns across your fleet in real time, and address them proactively before they spiral into an operational crisis. Put simply, a well-designed dashboard becomes the mission-critical control tower for your operations team and helps you elevate the reliability and performance of your fleet.

Hence, make it a point to develop a structured dashboard optimized for product reliability. If you don't, you will end up managing operational fires on a daily basis — and that brings us to the next point.

Lesson 05: Define Clear Emergency Response and Escalation Protocols

Failures in robotics, especially in the early stages, are inevitable. What differentiates successful companies in this arena from others is not just how reliable their product is, but how they respond when it does not perform reliably — when it goes down, stalls, or leads to a catastrophic failure. That naturally raises the question: how should a founder approach these operational fires?

From the get-go, as a founder, in addition to prioritizing product reliability, the second area of focus for you and your operations team should be to clearly define and adhere to emergency response and escalation protocols. These should be surgically defined for each and every client that you have onboarded and the ones that you will onboard. That naturally raises the next question: how should a founder define this protocol?

A Three-Pronged Approach to Building Your Emergency Response Protocol

We followed a three-pronged approach to define ours. To be specific, you need to define each of these three components with extreme precision. So, what are these?

1. Group Clients Strategically and Define Operational Ownership

First, you should be able to group your clients either by geography, by size, or by industry vertical — whichever is more practical in your case. This grouping of clients is intended to ensure that you split ownership for each of these groups equally across your employee base. This means that even if you don't have enough resources to dedicate one person per client, this grouping will allow you to distribute the workload evenly among your available operational talent.

2. Assign Two Cross-Trained Individuals per Client Group

Second, for each group of clients, you need to dedicate at least two individuals from your team. One must be an engineer, and the other should function more as an operations manager or field operator. Make no mistake: both should be cross-trained in case one of them is not available when such fires occur.

3. Surgically define Emergency Response Protocol Per Client

Once you have grouped your clients and allocated two dedicated individuals to handle emergencies for each group, the next step is to define the emergency response protocol. This should be agreed upon between you, your client, and your team. It should explicitly document how quickly the individual should respond to such incidents, how they should respond, which methods they should use to communicate the incident to the client and internal teams, and the corresponding priorities they should keep in mind — whether it's shutting down the system, cordoning off the physical space to prevent further damage, or engaging first responders. It should also include where and how the details of such incidents should be documented to enable structured post-mortem sessions.

These protocols should be defined and documented for every single client based on their needs and your ability to address those needs during an emergency. Without this level of precision in your response framework, you will lose clients in no time due to such incidents.

Why Surgical Precision in Protocols Directly Impacts Client Retention

You may ask why this is the case. The reason is that in robotics, while it is already extremely hard to secure a client in the first place, it is even harder to retain them in the long run. Emergency scenarios resulting from sensor failures, site abnormalities, or unaddressed edge cases can escalate in no time and place both you and your deal on the chopping block almost instantly. These protocols will not only help you reduce downtime and minimize risk but also reassure clients from the get-go that your system will be under control even during such failures. They will help you build and sustain trust in these relationships even through such emergency scenarios.

While it is important to define these emergency protocols with surgical precision, it is equally important to communicate them clearly to your clients — and that brings us to our next pointer.

Lesson 06: Why Bi-Monthly Client Feedback Sessions Are Non-Negotiable

It is hard to stress the importance of conducting one-on-one sessions with every client almost on a bi-monthly basis in the first few years of your startup journey. You may ask: what is the need for this session, and more importantly, what happens if you don't take these seriously?

The core objective is to ensure you are fully aware of your clients' needs and, most importantly, the areas where they are dissatisfied. No client will communicate these proactively unless you engage in serious, structured discussion sessions with them on a twice-monthly basis. If you don't conduct these sessions and don't uncover points of dissatisfaction in a proactive manner, make no mistake: your clients, in most cases, will not come to you with this feedback — and worse, may terminate your contract without giving any feedback at all.

Thus, if you are serious about retaining your clients, you need to ensure that your operations team members are able to establish these sessions with clients almost on a twice-a-month basis. This exercise will not only help build sustained trust with your clients but will also help you proactively identify risks that can put these deals on the termination pathway. Armed with these insights, your operations team can then work closely with the product development team to prioritize these areas in your roadmap and minimize churn.

If you are at the early stages of your company and have only a couple of clients, you can run these sessions yourself as the founder. If you are at a later stage, meaning you have at least half a dozen clients or more — you can assign an operations manager to handle this for a group of clients.

You may ask: what should be the structure and cadence for internal teams to communicate on these core areas of priority for the company? In simple terms, how can you channel such signals received by each of these internal teams and distribute them across the organization effectively?

That leads us to our next pointer.

Lesson 07: Hold a Monthly Cross-Functional Alignment Forum— or Watch Deals Die

Once a month, all critical teams — operations, engineering, sales, and marketing — must come together to align priorities, discuss areas of risk for each client, and recalibrate operational focus for the company as a whole. This structured forum will help you channel signals from all teams through the organization, minimize noise in your operations, and enable operational coherence across the organization. Without this alignment, teams will continue to operate in silos, unintentionally ignore early signs of risk for your existing clients and your relationships with them, thereby putting your revenue channels in high-risk mode and ultimately driving the trajectory of your startup toward a wall of no return. That naturally raises the question: what should be the structure of these sessions, and how often should these be conducted?

Weekly Team Cadence to Prevent Internal Drift

While the structure and frequency will vary depending on your scale of operation, an approach that worked well for us from the very beginning till the end—i.e., acquisition—is as follows. In our case, each team hosted a weekly cadence session that was primarily designed for team members belonging to that team. For example, the engineering team had a weekly standup on Wednesday, and the operations team had a weekly meeting on Thursday. Similarly, marketing and sales had their weekly forums.

Monthly 1:1 Customer-Facing Sessions to Spot Critical Signals

In addition to these internal meetings, each of these teams interfaced with the customer at least on a monthly basis. Sales teams paired with operations and attended on-site sessions with customers to better understand their needs. The engineering team did the same with the operations team to better understand both the challenges of expanding our service at a customer site and the corresponding safety and reliability risks cited by customers.

Structured Monthly Town Halls & Boardroom Sessions

Insights from both these sessions—i.e., the internal team meetings and their corresponding client-facing external meetings—were collated at the end of each month and brought together by the captains of these teams to the founder. The founders then organized these insights, prioritized the areas of focus for all teams, and conducted a monthly company-wide session that was surgically crafted and centered on select priorities. You may ask: what were these areas of focus for these monthly sessions?

The Four Non-Negotiables for Monthly Town Halls

The four non-negotiable areas of focus for monthly sessions included: client-specific wins—i.e., what each of these clients was especially valuing about our service; client-specific losses—what they absolutely hated about our product and service; component-specific failures—which of our components, both hardware and software, were consuming a lot of our engineering and operations time and hence were not reliable and needed to be replaced; high-risk incidents—as signals to sales and operations teams to tailor the type of client sites that our robots were most suited for and say no to deals that may not be aligned to the risk profile of our product.

This one-hour session, organized by our founders each month with near-monastic devotion, brought an extraordinary level of clarity, operational coherence, and urgency across all teams. Make no mistake, each of these sessions was recorded so that if an employee was leading a deployment in the field and unable to attend, they could still stay aligned with the updated priorities conveyed to all employees on a monthly basis. Last but not least—as a founder, it goes without saying—these sessions also help you derive core insights for your boardroom meetings, ensuring your investors are well informed about where you are as a company, what the current top 2–3 priorities are and why, and how you plan to address them.

If you truly want to succeed as a robotics founder, make this one of your top five priorities.

That naturally leads to the next question: how do you centralize the insights gathered through these sessions and channelize them into your long-term operational protocols? That's where the next pointer becomes critical.

Lesson 08: Build a Centralized Knowledge Repository — or Repeat the Same Failures Forever

While it is understood that a robotics team should maintain a centralized knowledge portal that includes client-specific SOPs, incident logs, deployment guides, and more, one critical mistake founders make is failing to dedicate resources from each team to take ownership of specific sections of this knowledge repository. This is important because no matter how well drafted this portal is, at the end of the day, it's a living document. After every customer deployment, it must be updated to mitigate risks and improve the overall quality of future deployments. This means someone needs to update it continuously and refine it based on the latest operational realities. This is where you need to define a clear protocol for how this portal will be updated.

A Two-Phase Protocol to Structure Your Knowledge Repository

You might ask: how should a founder define this protocol? While there is no right or wrong answer here, in our case we adopted a two-phased approach.

In Phase I, each month, an operations manager would connect with a group of clients assigned to them in a given geography and collate insights post-deployment around what worked and what needed to be changed for that group of clients. In Phase II, they would organize and share these insights with an engineer who would update the portal accordingly after each deployment. The inputs here were collected by the operations manager post every deployment for their set of clients and then channeled to engineering resources, who updated the documentation in alignment with evolving protocols. This ensured operational protocols around our product and service were surgically updated, remained centralized, and were available to all teams at all times.

Now, as we are talking about protocols and knowledge-sharing process, it goes without saying that in robotics you will benefit more from system thinkers than from specialists.

Why System Thinkers Matter More Than Specialists

A robot, at the end of the day, is made of a hundred different subsystems—from sensors to cameras and from batteries to motors. Your centralized repository will include lessons learned and protocols that clearly define how your demos and new field deployments should be conducted, along with the corresponding areas of risk that the person on site should keep in mind. This means the centralized repository will contain vital lessons that cover the full hundred yards of robotics—from motors to sensors, cameras to batteries, and networking modules. Hence your engineering team should predominantly be made of system thinkers and not specialists. You will need folks who can understand each of these systems inside out and can resolve any issue end to end. This is where the next lesson comes in handy.

Lesson 09: Embed Systems Thinking and Ownership at Every Level

Just like a cloud system is a network of servers connected to each other through optical fibers that run beneath our oceans and land, a robot, at the end of the day, is a network of subsystems that are connected to each other and responsible for capturing data and translating that into action. Some of these subsystems—such as sensors and cameras—capture raw data, while others—such as CPUs and GPUs—process this data and finalize actions for the robot. These actions are then executed by yet another set of subsystems such as batteries, motors, tablet screens, robot arms, and more. The challenge here is that these subsystems do not work in isolation. They work in unison, and therefore every decision a team member makes when building a feature or fixing a robot carries downstream risk.

Build a Team of System Thinkers to Better Manage Risk

Your team must therefore think in systems, not tasks. Each team member should be able to decode the implications of altering one subsystem on another. They should understand the repercussions of modifying the code module responsible for handling video feeds from the camera on the robot's perception and navigation system. Similarly, they should also understand the downstream impact of altering code responsible for the robot's networking capabilities on its overall reliability. Put simply, each team member should develop cross-functional expertise. This is not an optional need in robotics—it's a must.

You may ask: why is this a must, and more importantly, how do you help these folks develop cross-functional expertise?

To better understand the *why*, it goes without saying that a robotics engineer by default may be reasonably good with sensors and planners but may have no understanding of the networking and telecom module on the robot. This means if a robot goes down due to network outage, this engineer will not be in a position to help the company. Hence, it is critical that they develop enough expertise to at least debug and identify root causes of failure in these modules. The same applies the other way around: an engineer whose primary expertise revolves around wireless networking should develop foundational understanding of the robotics stack as well. These team members should think in terms of systems and downstream risks. They should develop the habit of asking, *"If this breaks, what else breaks with it?"*

This leads to the next question: how can you as a founder help team members develop cross-functional expertise across your organization?

Monthly Charter to Build Cross-Functional Expertise Across the Team

First, it goes without saying—hire system thinkers from the get-go.

Second, if you have already hired specialists and now need to help them develop holistic understanding of your robot, have each specialist hold a monthly session focused on distilling the most important areas within their specialization into a two-hour session for all team members. In simple worlds, have them conduct a monthly knowledge-sharing session and document this information. This will help them refine their own understanding of their domain as well as simplify this information for those who are new to that area of expertise.

Your goal at the end of the day as a founder is to ensure that every individual on your team—including your operations managers—thinks like an owner, not merely an operator. This mindset reduces error propagation and increases accountability. Without it, problems quickly become "someone else's responsibility," and systemic failures will eat your financial runway in no time. It goes without saying that, as a founder, you must incubate this philosophy within your team from day one. That leads us to our next pointer.

Lesson 10: Deep Founder Engagement in Core Execution Zones Is Non-Negotiable

As a robotics founder, it becomes extremely important that you remain deeply engaged across all major verticals of your company. Your goal here is to establish an almost weekly cadence with your sales, engineering, operations, and customer success teams, track major wins and most importantly core areas of risks. Delegation without strategic vigilance across these fronts, is a recipe for disaster, to say the least. You may ask, why is that the case?

The Hidden Cost of Detaching Too Early

Operational fires in robotics are inevitable at the early stage, and the best way to deal with them is to be extra prepared from day one. For you to be well prepared as a founder, you need to be deeply engaged across all these areas so you have clear data points on what is working in your product and protocols, what is not, and how you can go about fixing it before it turns into another fire. One of the major mistakes first-time founders make in this arena is that they detach too early from these teams and over-rely on delegated modes of operation. This weakens the feedback loop, and you end up losing critical signals until they turn into an operational crisis.

So how should you conduct these sessions, and more importantly, what should you collate?

How to Structure Your Weekly Sessions

As a founder, during each of these sessions from each of your team leads, here is a list of insights you must collate. This is only Step 01:

1. What are we doing today in this area that is or will help us in the next 30 days to land a new client—and how?
2. What are we doing today in this area that is or will help us in the next 30 days to save an existing client—and how?
3. What are we not doing today in this area that, if we start doing in the next 30 days, will help us land a new client—and how?
4. What are we not doing today in this area that, if we start doing in the next 30 days, will help us save an existing client—and how?

Beyond these four questions, the rest is all standard metrics, product reliability risks, risk mitigation, and other operational details. Those are important, but nothing is as important as understanding how your company as a whole is marching toward growth. And growth here ultimately boils down to your ability to drive adoption at an exponential rate while minimizing risks. In simple words, as a founder, your ability to consistently generate revenue for your employees and investors while creating value for your customers is the ultimate determinant of your success.

Be vigilant on this front.

That completes some of the core lessons across all operational fronts that you must keep in mind as a robotics founder. Now that you have read this book, in the next chapter, you will learn the next steps you can take to strategically apply the lessons and frameworks discussed in this book.

Chapter 25: How Founders & Engineers Can Apply This Book

Thank you so much for reading this book. Whether you are a founder building your robotics startup or an engineer trying to break into this field, here are the six steps you can take to kickstart the Boring Sage process and immediately benefit from the lessons and frameworks shared throughout this book.

Part A: 6 Steps for Founders

1. **Define Your Client Strategy & Field-Deployment Model**
   Read Section II and deeply understand the lessons on structuring your client portfolio (Fortune 500 vs mid-sized), refining your approach to working with these clients, evaluating conferences and trade shows to demonstrate your product, and understanding the technology and site-specific risks that will emerge once you start deploying robots in the field.
2. **Apply the Engineering Playbook Built for Your Robot Class**
   If you are building a humanoid robotics company, read Section III and understand how Tesla approached several of the challenges you will face as you begin building one. If you are building mobile robots, quadrupeds, or robot

arms, focus on Section IV, as it covers the exact approaches adopted by companies like Amazon and Boston Dynamics to build these classes of robots and the corresponding lessons that come with them.

3. **Identify the Cross-Functional Risks That Break Robotics Startups**
   Read Chapters 21–24 from Section VI. They provide an in-depth overview of the core risks you will encounter across your sales, marketing, operations, engineering, and talent management teams as you build your robotics startup. More importantly, they will also equip you with the corresponding solutions you can apply to address these risks proactively.
4. **Build a Dedicated Risk Map for Each Team You Can Act On**
   As you work through these sections, create a running list of risks relevant to each team. I recommend maintaining four separate A4 pages—one each for sales, engineering, marketing, and field operations—and capturing the applicable risks and lessons for every group. This will help you clearly visualize the most vulnerable areas of your organization and equip you with the corresponding lessons to address them.
5. **Assign Owners and Establish Quarterly Accountability Channels**
   Once your risks are organized across these four A4 pages, you will have clear, actionable insights to share with each team. Use these insights to help your team proactively address each of those bottlenecks. Keep this book and these four pages as a quarterly calibration tool across

sales, engineering, marketing, and operations. This ensures you remain aligned and ahead of emerging issues.

6. **Have Your Team Read This Book & Align on One Framework**
   Ensure every team member understands the lessons in this book and the context behind them. They should know why each lesson matters, how to identify risks early, and how to respond with clarity. A shared and standardized framework for each area of your company—sales, engineering, marketing, and operations—especially the one distilled in the chapters in Section VI, creates common ground for your team to address challenges in a calm, collaborative, consistent, and solution-oriented manner. It ensures every team member is aware of the repercussions of unresolved risks, drives urgency, and brings cosmic clarity to their approach.

Part B: 6 Steps for Engineers

1. **Learn the Real Risks of Working in the Field**
   Begin by studying the risks you will encounter once you start working directly with clients in the field—demonstrating your robot-led product or service at their offices, conferences, and trade shows. These include technology failures, safety issues, security gaps, navigation anomalies, client expectations, and site-specific challenges. Understanding these early will help you build the right mental model for this industry.

2. **Internalize the Core Operating Principles for Managing Risk**
   Study Section I and equip yourself with the core principles required to handle those risks, i.e., operational fires that will naturally arise once you start working with clients in the field. It will also give you historical context on the trajectory of a few consumer robotics companies and will help you understand why choosing the right few companies will matter more than chasing many when you begin your job search.
3. **Select Your Robot Class and Master Its Engineering Playbook**
   Based on the companies you plan to approach, finalize the robot class most relevant to your goals—humanoids, robot arms, or mobile warehouse robots. Then study the corresponding engineering playbooks decoded in Sections III and IV. Here you will learn how top players build these systems and tackle the technical and operational risks associated with each class.
4. **Study the Career Traps You Will Encounter Early in Your Journey**
   Every career trajectory has hidden cliffs. Robotics engineering is no different. Knowing these early will help you navigate them safely and save you years of misdirected effort. Head to Section V and study Chapters 18 and 19. These chapters uncover the traps inherent in this career path and the ones you will face as you begin your job search.

5. **Use the Structured System to Target the Right Roles**
   Study Chapter 20 in Section V. Here, you will learn a step-by-step framework to identify roles and startups aligned with your career mission—and a clear, structured method for landing them with confidence. This system will help you consistently identify and convert opportunities that truly align with your long-term goals in this space.
6. **Apply the Framework, Approach the Right Companies, and Demonstrate Your Value**
   Using the lessons and frameworks you've studied in Chapter 20, identify the companies and roles aligned with your mission. Before you approach them, refine your resume based on the class of robot you have studied along with the corresponding case studies of top players in the field. Add these as independent project work to your resume, demonstrating that you have taken the time to study the challenges faced by leading companies in this space and the approaches they use to address them.
   Next, approach founders and engineering teams of companies you filtered using the method outlined in this section. During interviews, articulate the risks, insights, and frameworks you've learned. Your objective is to create a win-win—demonstrating that you understand the challenges robotics companies face across verticals and that you bring a structured, proactive approach to address them.
   This positions you as more than a technically capable engineer. It positions you as someone who understands how this industry actually works—shaped by real risks,

proven frameworks, and the operating playbooks of the world's top robotics teams.

You're Not Alone on This Path

Building a robotics company or landing a job in the right one—both are equally hard. I know that many of you reading this book are either engineers trying to break into this field or founders trying to bring your company and its mission to the global stage. In either case, I understand that this journey is not easy. The winds of distraction and the cliffs of failure can be felt with every step you take here. That said, you are not alone on this path.

1. **Contact my team:** Reach out to my team using the chatbox located on https://boringsage.com if you ever have a question about anything. If not myself, someone from our team will address your question.
2. **Explore our resources:** We regularly share deep dives on career trajectories in this space, startup lessons, and technology breakthroughs on our blog. Based on the problem you are trying to solve, in addition to this book, you will find these resources valuable. You can explore them here: https://boringsage.com/blog
3. **Join our community:** We have a community of robotics professionals who have signed up for our monthly newsletter. We regularly post updates on the resources we share on our blog, upcoming launches of our books, products, and courses. You can sign up for free here: https://www.boringsage.com/pulse

Thank You

At Boring Sage, our mission is to bring deep professional expertise to those who don't have access to it firsthand. From equipping founders with hard-earned lessons from the field to helping engineers approach their careers with a structured framework, our goal is to enable you to navigate the risks inherent on this path proactively—so you can bring the next generation of robots to life with peace of mind and address emerging challenges in society.

It is my hope that the lessons and frameworks discussed here will be incredibly valuable on your journey. I wish you the very best as you carry these lessons into the field, and I look forward to seeing your robots in action in the real world.

Image Credits

All diagrams, schematics, and illustrations in this book were created by the author unless otherwise stated. Portions of select figures include images reproduced or excerpted from publicly available materials from the following organizations and sources, used solely for educational and illustrative purposes:

- **Tesla, Inc.** — Tesla AI Day technical presentations; the Tesla "We, Robot" event; the Tesla YouTube channel (including "Optimus - Gen 2"); and other publicly released Tesla media assets.
- **Boston Dynamics** — Official product demonstrations on the Boston Dynamics YouTube channel; official Boston Dynamics blog articles (including reinforcement-learning research); Boston Dynamics product and case-study pages (including the Stretch platform and the DHL deployment); research publications; and publicly released media materials.
- **Amazon Robotics / Amazon, Inc.** — About Amazon corporate news and media resources; Amazon News YouTube channel; Amazon Science research publications; and publicly released warehouse automation and robotics media assets.

References

[1] A. E. LeBouthillier, "W. Grey Walter and his Turtle Robots," The Robot Builder, vol. 11, no. 5, pp. 1–3, May 1999. [Online]. Available: https://home.csulb.edu/~wmartinz/sites/default/files/newsletter/mar99.pdf

[2] B. Malone, "George Devol: A life devoted to invention, and robots," IEEE Spectrum, Sep. 26, 2011. [Online]. Available: https://spectrum.ieee.org/george-devol-a-life-devoted-to-invention-and-robots

[3] N. J. Nilsson, *Shakey the Robot*, Tech. Note 323, Artificial Intelligence Center, Computer Science and Technology Division, Apr. 1984. Approved by D. L. Nielson. [Online]. Available: https://ai.stanford.edu/~nilsson/OnlinePubs-Nils/shakey-the-robot.pdf

[4] J. Leber, "Willow Garage won't do research anymore, but it'll sell you a robot," *MIT Technology Review*, Feb. 12, 2013. [Online]. Available: https://www.technologyreview.com/2013/02/12/17100/willow-garage-wont-do-research-anymore-but-itll-sell-you-a-robot/

[5] K. Wiggers, "Robotics startup Anki shuts down after burning through almost $200 million," *VentureBeat*, Apr. 29, 2019. [Online]. Available: https://venturebeat.com/ai/robotics-startup-anki-shuts-down-after-burning-through-almost-200-million

[6] E. Ackerman, "Mayfield Robotics cancels Kuri social home robot," *IEEE Spectrum*, Jul. 25, 2018. [Online]. Available: https://spectrum.ieee.org/mayfield-robotics-cancels-kuri-social-home-robot

[7] H. M. Varghese and J. Babu, "Amazon to wind down its Astro for Business security robot," *Reuters*, Jul. 3, 2024. [Online]. Available: https://www.reuters.com/technology/amazon-wind-down-its-astro-business-security-robot-2024-07-03

[8] E. Ackerman, "Zebra Technologies to acquire Fetch Robotics for $305 million," *IEEE Spectrum*, Robotics News, Jul. 1, 2021. [Online]. Available: https://spectrum.ieee.org/zebra-technologies-acquire-fetch-robotics

[9] A. Barr, "Amazon.com to buy Kiva Systems for $775 million," *Reuters*, Mar. 19, 2012. [Online]. Available: https://www.reuters.com/article/business/amazoncom-to-buy-kiva-systems-for-775-million-idUSBRE82I117/

[10] A. Hamer, "How Many Megapixels Is the Human Eye?," *Discovery (Curiosity.com)*, Aug. 1, 2019. [Online]. Available: https://www.discovery.com/science/mexapixels-in-human-eye

[11] C. Caruso, "A New Field of Neuroscience Aims to Map Connections in the Brain," *Harvard Medical School News*, Jan. 19, 2023. [Online]. Available: https://hms.harvard.edu/news/new-field-neuroscience-aims-map-connections-brain

[12] C. Chen, "China's electric vehicle giants are betting big on humanoid robots," *MIT Technology Review*, Feb. 14, 2025. [Online]. Available: https://www.technologyreview.com/2025/02/14/1111920/chinas-electric-vehicle-giants-pivot-humanoid-robots/

[13] W.-Y. Lee, "Foxconn to deploy humanoid robots at Houston AI server plant," *Reuters*, Oct. 28, 2025. [Online]. Available: https://www.reuters.com/business/foxconn-deploy-humanoid-robots-houston-ai-server-plant-2025-10-29/

[14] *F.02 Contributed to the Production of 30,000 Cars at BMW*, Figure.ai, Nov. 19, 2025. [Online]. Available: https://www.figure.ai/news/production-at-bmw

[15] Hyundai Motor Company, "Uniting humans and robots – Hyundai Motor Group and Boston Dynamics' vision for the future of work and HR evolution," *Hyundai Worldwide – Newsroom*, Jul. 14, 2025. [Online]. Available: https://www.hyundai.com/worldwide/en/newsroom/detail/uniting-humans-and-robots--hyundai-motor-group-and-boston-dynamics-vision-for-the-future-of-work-and-hr-evolution-0000000983

[16] I. Fujino, "China's BYD, Nio do more with AI-powered humanoid factory robots," *Nikkei Asia*, Dec. 19, 2024. [Online]. Available: https://asia.nikkei.com/business/china-tech/china-s-byd-nio-do-more-with-ai-powered-humanoid-factory-robots

[17] Porsche Consulting GmbH, "The 'real' colleague," *Porsche Consulting – The Magazine*, Dec. 11, 2025. [Online]. Available: https://newsroom.porsche.com/en/2025/company/porsche-consulting-the-real-colleague-41063.html

[18] D. Milmo, "Amazon 'testing humanoid robots to deliver packages'," *The Guardian*, Jun. 5, 2025. [Online]. Available: https://www.theguardian.com/technology/2025/jun/05/amazon-testing-humanoid-robots-to-deliver-packages

[19] N. Khan, "Tesla's Optimus robots to enter mass production in 2025, aims for 100,000 by 2026," *Yahoo Finance*, Jan. 10, 2025. [Online]. Available: https://finance.yahoo.com/news/teslas-optimus-robots-enter-mass-101736202.html

[20] Business Wire, "Global humanoid robots market report 2026–2040: Humanoid robot pricing drops below $10,000 as market expansion accelerates," *The AI Journal*, Oct. 29, 2025. [Online]. Available: https://aijourn.com/global-humanoid-robots-market-report-2026-2040-humanoid-robot-pricing-drops-below-10000-as-market-expansion-accelerates-researchandmarkets-com/

[21] Agility Robotics, "RoboFab: The first humanoid robot factory," *Agility Robotics*, 2025. [Online]. Available: https://www.agilityrobotics.com/about/robofab

[22] J. Biba, "Tesla's Optimus robot: Uses and release timeline," *Built In*, updated Nov. 17, 2025. [Online]. Available: https://builtin.com/robotics/tesla-robot

[23] K. Walia, "A humanoid robot is now on sale for under US$6,000 – what can you do with it?," *The Conversation*, Aug. 18, 2025. [Online]. Available: https://theconversation.com/a-humanoid-robot-is-now-on-sale-for-under-us-6-000-what-can-you-do-with-it-262183

[24] C. Stokel-Walker, "Do robots dream of perfect hands? Why we're so obsessed with giving robots human digits," *Fast Company*, Oct. 9, 2025. [Online]. Available: https://www.fastcompany.com/91418249/do-robots-dream-of-perfect-hands-optimus-musk-robot

[25] Tesla, "Tesla AI Day 2022," YouTube video, Sep. 30, 2022. [Online]. Available: https://www.youtube.com/watch?v=ODSJsviD_SU

[26] Amazon Staff, "Why Amazon warehouses are called fulfilment centres and how many are in the UK," *Amazon News*, n.d. [Online]. Available: https://www.aboutamazon.co.uk/news/operations/why-amazon-warehouses-are-called-fulfilment-centres

[27] T. Greenawalt, "Amazon has more than 1 million robots that sort, lift, and carry packages—see them in action," *Amazon News*, updated Oct. 22, 2025. [Online]. Available: https://www.aboutamazon.com/news/operations/amazon-robotics-robots-fulfillment-center

[28] Capital One Shopping, "Amazon logistics statistics," *Capital One Shopping Reports*, updated Jul. 23, 2025. [Online]. Available: https://capitaloneshopping.com/research/amazon-logistics-statistics/

[29] A. Davies, "Introducing Vulcan: Amazon's first robot with a sense of touch," *Amazon News*, May 7, 2025. [Online]. Available: https://www.aboutamazon.com/news/operations/amazon-vulcan-robot-pick-stow-touch

[30] Amazon Staff, "Meet Proteus: I'm Amazon's first autonomous robot," *Amazon News*, Oct. 11, 2024. [Online]. Available: https://www.aboutamazon.com/stories/amazon-robotics-autonomous-robot-proteus-warehouse-packages

[31] Amazon Staff, "How does your Amazon order get to your doorstep? Watch the journey of a package," *Amazon News*, updated Apr. 26, 2024. [Online]. Available: https://www.aboutamazon.com/news/operations/how-do-amazon-packages-get-delivered

[32] Amazon Staff, "Employees," *Amazon Sustainability*, n.d. [Online]. Available: https://sustainability.aboutamazon.com/employees

[33] J. Durham, "Amazon builds first foundation model for multirobot coordination," Amazon Science, Aug. 11, 2025. [Online]. Available: https://www.amazon.science/blog/amazon-builds-first-foundation-model-for-multirobot-coordination

[34] Amazon Staff, "Amazon announces 3 AI-powered innovations to get packages to customers faster," Amazon News, Jun. 11, 2025. [Online]. Available:

https://www.aboutamazon.com/news/operations/amazon-ai-innovations-delivery-forecasting-robotics

[35] S. Dresser, "Amazon announces 2 new ways it's using robots to assist employees and deliver for customers," *Amazon News*, Oct. 18, 2023. [Online]. Available: https://www.aboutamazon.com/news/operations/amazon-introduces-new-robotics-solutions

[36] R. Drew, "Amazon prepares to test humanoid robots for delivering packages," *The Information*, Jun. 4, 2025. [Online]. Available: https://www.theinformation.com/articles/amazon-prepares-test-humanoid-robots-delivering-packages

[37] T. Greenawalt, "Amazon's newest AI and robotics systems are empowering employees and speeding up delivery," *Amazon News*, Oct. 22, 2025. [Online]. Available: https://www.aboutamazon.com/news/operations/amazon-delivering-future-2025-online-shopping-speed-delivery

[38] E. Ackerman, "Boston Dynamics' founder on the future of robotics," *IEEE Spectrum*, Jul. 31, 2023. [Online]. Available: https://spectrum.ieee.org/boston-dynamics

[39] E. Ackerman, "Hyundai buys Boston Dynamics for nearly $1 billion. Now what?" *IEEE Spectrum*, Dec. 11, 2020. [Online]. Available: https://spectrum.ieee.org/hyundai-buys-boston-dynamics

[40] DHL Group, "DHL Group signs MOU with Boston Dynamics for additional 1,000-robot deployment and accelerates cross-business automation strategy," Press Release, May 13, 2025. [Online]. Available: https://group.dhl.com/en/media-relations/press-releases/2025/dhl-group-signs-mou-with-boston-dynamics-and-accelerates-cross-business-automation-strategy.html

[41] Boston Dynamics, "Case studies," *Boston Dynamics*, [Online]. Available: https://bostondynamics.com/case-studies/

[42] U.S. Department of Labor, Mine Safety and Health Administration, "Carbon Monoxide and Mining," *Health Hazard Information Card* HH-28, May 2007. [Online]. Available:

https://www.msha.gov/sites/default/files/Alerts%20and%20Hazards/HH%20Cards/HH-28%20Carbon%20Monoxide%20&%20Mining.pdf

[43] M. Hall, "Researchers compile first-ever global record of mining land use," *Mining Technology*, GlobalData, Sept. 22, 2020. [Online]. Available: https://www.mining-technology.com/news/researchers-compile-first-ever-global-record-of-mining-land-use/

[44] InformedHealth.org, *"In brief: How do lungs work?"* National Center for Biotechnology Information (NCBI), U.S. National Library of Medicine, Apr. 18, 2023. [Online]. Available: https://www.ncbi.nlm.nih.gov/books/NBK401240/

[45] Boston Dynamics, "Boston Dynamics' legacy robots," Boston Dynamics, 2025. [Online]. Available: https://bostondynamics.com/legacy/

[46] Boston Dynamics, "Spot – The Agile Mobile Robot," *Boston Dynamics*, 2025. [Online]. Available: https://bostondynamics.com/products/spot/

[47] Boston Dynamics, "3 Ways Robots See the World," *Boston Dynamics Blog*, Trends in Robotics, 2025. [Online]. Available: https://bostondynamics.com/blog/3-ways-robots-see-the-world/

[48] Boston Dynamics, "Starting on the right foot with reinforcement learning," *Boston Dynamics Blog*, Innovation, 2025. [Online]. Available: https://bostondynamics.com/blog/starting-on-the-right-foot-with-reinforcement-learning/

[49] Boston Dynamics, "Spot watches its step," YouTube video, Nov. 14, 2024. [Online]. Available: https://www.youtube.com/watch?v=YD9EaS3VRbc

[50] Boston Dynamics, "How Spot Moves," *Boston Dynamics Support Center*, Boston Dynamics, Boston, MA, USA, 2025. [Online]. Available: https://support.bostondynamics.com/s/article/How-Spot-Moves-49929

[51] Boston Dynamics, *Stretch Brochure 2025*, Boston Dynamics, 2024. [Online]. Available: https://bostondynamics.com/wp-content/uploads/2024/01/Stretch-Brochure-2025-compressed.pdf

[52] Boston Dynamics, "From idea to application with Stretch," *Boston Dynamics Blog*, Warehouse Robotics, 2025. [Online]. Available: https://bostondynamics.com/blog/idea-to-application-with-stretch/

[53] DHL Group, "Freight shipping options: DHL Global Forwarding and DHL Freight," DHL, 2025. [Online]. Available: https://www.dhl.com/ba-en/home/ship/freight-shipping-options.html

[54] DHL Supply Chain, "DHL Supply Chain achieves first commercial deployment of Boston Dynamics' Stretch™ robot to unload trailers and containers in the warehouse," Press Release, Westerville, OH, USA, Jan. 31, 2023. [Online]. Available: https://www.dhl.com/us-en/home/press/press-archive/2023/dhl-supply-chain-achieves-first-commercial-deployment-of-boston-dynamics-stretch-robot.html

[55] Boston Dynamics, *"Stretch at DHL,"* Case Study, Boston Dynamics, 2025. [Online]. Available: https://bostondynamics.com/case-studies/stretch-at-dhl/

[56] G. Aylward and L. Miller, "What it takes to pick a million boxes," *Boston Dynamics Blog*, Aug. 2023. [Online]. Available: https://bostondynamics.com/blog/what-it-takes-to-pick-a-million-boxes/

[57] D. Levine, "China may have more engineers, but it still lacks a culture of innovation," *The Hill*, Aug. 12, 2025. [Online]. Available: https://thehill.com/opinion/international/5447381-china-may-have-more-engineers-but-it-still-lacks-a-culture-of-innovation/

[58] N. Sharma, "Why are so many of India's 1.5 million fresh engineers every year unemployable?" *Forbes India*, Nov. 13, 2025. [Online]. Available: https://www.forbesindia.com/article/upfront/column/why-are-so-many-of-indias-1-5-million-fresh-engineers-every-year-unemployable/2988705/1

[59] A. Kodey, J. Bedard, J. Nipper, N. Post, S. Lovett, and A. Negreros, "The US needs more engineers. What's the solution?" *Boston Consulting Group*, Dec. 13, 2023. [Online]. Available: https://www.bcg.com/publications/2023/addressing-the-engineering-talent-shortage

[60] J. Stidd, "Raising on your terms: Lessons from robotics founders who took control of their capital strategy," *DealMaker Webinar*, Jun. 24, 2025. [Online]. Available: https://www.dealmaker.tech/content/raising-on-your-terms-lessons-from-robotics-founders-who-took-control-of-their-capital-strategy

[61] DealMaker, "Robotics Venture Funding: Navigating Capital Raising in 2025," *DealMaker Blog*, Apr. 14, 2025. [Online]. Available: https://www.dealmaker.tech/content/robotics-venture-funding-navigating-capital-raising-in-2025

[62] Fieldwork Robotics, "Fieldwork closes successful funding round with Seedrs," *Fieldwork Robotics*, 2025. [Online]. Available: https://fieldworkrobotics.com/fieldwork-closes-funding-round-2/

[63] Fieldwork Robotics, "Fieldwork announces £1.5m investment from Elbow Beach Capital," *Fieldwork Robotics*, Cambridge, U.K., 2025. [Online]. Available: https://fieldworkrobotics.com/example-article-3/

[64] Zipline, "Zipline announces new funding," *Zipline*, Team Zipline, 2021. [Online]. Available: https://www.zipline.com/newsroom/news/announcements/zipline-announces-new-funding

[65] Global Conference Alliance, "How much does it cost to exhibit at a conference?" *Global Conference Alliance*, Oct. 22, 2025. [Online]. Available: https://globalconference.ca/how-much-does-it-cost-to-exhibit-at-a-conference/

Notes on Sources, Author Experience, and Educational Use

This book brings together lessons learned from the field with independent analysis of publicly available material. Certain chapters draw on the author's first-hand professional experience in building, deploying, and working with robots in real-world environments, and reflect practical observations and insights developed through that work.

Other chapters examine the approaches adopted by leading robotics companies and present technical descriptions derived from publicly available sources, including company disclosures, technical presentations, product documentation, academic publications, and industry analyses. These materials have been independently interpreted, synthesised, and contextualised by the author. The resulting analysis represents the author's understanding of the subject matter and does not constitute official documentation, endorsement, or representation by the companies or organisations referenced.

The content of this book is intended solely for educational and informational purposes. It is not intended to provide professional, engineering, legal, or commercial advice, and should not be relied upon as a substitute for independent evaluation, validation, testing, or consultation with qualified professionals.

About the Author

Prathamesh Khedekar is the founder of Boring Sage, an ed-tech company focused on bringing deep professional expertise in emerging technologies to engineers and founders. He has built mission-critical systems for the governments of the U.S., U.K., Norway, and the Netherlands, and autonomous robotic systems for Fortune 100 companies globally. An Einstein Visa recipient, his work has been featured in The San Diego Union-Tribune, Security Magazine, and American Banker. He writes and teaches at the intersection of emerging technologies and business applications, helping founders and engineers bring next-generation products to life.

Printed by Libri Plureos GmbH in Hamburg,
Germany

9 798994 296608